CARTOON FRIENDS OF THE BABY BOOM ERA

A PICTORIAL PRICE GUIDE BY BILL BRUEGMAN

EDITED BY
JOANNE M. BRUEGMAN

CAP'N PENNY PRODUCTIONS, INC.
AKRON, OHIO

NOTICE

The prices in this book are approximations: neither the author nor the publisher may be held responsible for any losses that may occur through the use of this book in the purchase or sale of items.

First Printing

Published in Akron, Ohio by Cap'n Penny Productions

Photography by Bill Bruegman
Front cover photography by Tom Bosshard of Pro Color, Akron, Ohio.
Layout by Cap'n Penny Productions, Akron, Ohio.

Bruegman, William R. III
Cartoon Friends of the Baby Boomer Era, A Pictorial Price Guide/by Bill Bruegman
ISBN 0-9632637-3-0

Manufactured in the United States of America

ACKNOWLEDGEMENTS

The following people proved invaluable to me in this endeavor by providing me with guidance and advice based on their various skills and their friendship: Tom Bosshard of Pro Color, Ken Sharp, Lisa Meilander, Paul Matheis, Norm Vigue, and Mike Melito. I especially give my appreciation to Hal Lifson for contributing his writing skills to compose the introduction and the articles.

Finally, I lovingly acknowledge my wife Joanne, without whose support and long hours of typesetting, proofreading, and editing this book would not have been possible.

TABLE OF CONTENTS

ITEMS NOT COVERED IN THIS BOOK

Items logged in Cartoon Friends of the Baby Boom Era are divided into nineteen general topics. Naturally, it is impossible to include every toy produced between 1946-1969. To make Cartoon Friends concise and more valuable as a resource, some forms of collectibles are not covered.

The rule for exclusion is simple: You won't find it here if there is already a definitive, specialized source. The obvious example is comic books; many fine reference books and price guides already exist on the subject.

Collectibles not included in Cartoon Friends of the Baby Boom Era are:

* Comic Books
* Movie and TV stills
* Posters
* TV Guides

REGARDING PRICES IN THIS BOOK

The prices in Cartoon Friends of the Baby Boom Era have been compiled from a combination of factors, including the author's experience, sale prices of specific items from the Toy Scouts, Inc. Mail Order Catalog, typical prices at collectible shows, dealer price lists, ads appearing in collectibles periodicals, auction results, and extensive correspondence with collectors involved in trading, buying, and selling.

The prices quoted in this book are for Near Mint or Mint items, still in their original packaging, if any, and are collector, or retail, prices. The approximate value of a toy not in its original packaging, but still in very clean and near mint condition would be one half the value of the lowest end of the price range.

There will generally be a high and low range for the prices quoted in this book. This reflects geographical price variations of the marketplace. Ever-increasing demand for scarcer items or a newly discovered warehouse find of items previously difficult to obtain may have an upward or downward effect on prices, therefore it is imperative to keep in mind that this book serves as a guide in an appraisal and does not set the market value. Collectibles, like the stock market, work on the premise that the item is only worth what someone is willing to pay for it. The broader your clientele, the better chance you have for receiving a higher price.

Remember, selling is an art. In fact, knowing to whom to sell your collectible to is worth at least 50% of its value!

GRADING TOYS

Collecting toys requires a particular state of mind. It's a curious brew of nostalgia, aesthetics and investment strategy. Because there are so many diverse motives for collecting, one or more of the triad mind-set can dominate. Some collectors want an item because they had one like it as a kid. Others may stumble across a piece and think it looks "cool" or visually pleasing. Still others purchase collectibles because they're sure the items will appreciate quickly.

No matter what an individual's reason is for collecting, everyone agrees that the most desirable pieces are those in near-perfect, or "mint", condition.

Unfortunately, items turn up more frequently in less-than-mint condition than they do in factory-fresh shape. Some of the paint may have worn off, a decal may have been scraped away, or the nose on a character's face might have been broken. Moving parts may no longer move. Detachable accessories could be missing. In other words, at the opposite end of the scale are the real beaters.

Between the pristine mint examples of an item and those that are pretty well trashed are literally thousands of possibilities. Beyond the two extremes, it is rare that two collectors completely agree on the state of an item. A flaw that may seem insignificant to one person may make it totally unacceptable to another. Because of the wide-spread difference of opinion on condition, there has never been--and more than likely will never be--a universal standard of grading accepted by all buyers and sellers. However, general terms have been coined and adopted throughout the market to help evaluate items for the pruposes of trade, resale or appraisal. Many of these terms are commonly used in other types of collecting (i.e. cards, comics, and posters.)

The following scale is used to describe merchandise for sale in the mail order catalog of Toy Scouts (available for $3.00 from Toy Scouts, Inc., 137 Casterton Avenue, Akron, OH 44303).

MINT: Flawless; like new and, in most cases, unused.

NEAR MINT (NM): Only slight detectable wear, usually very minor. Overall appearance is like new in every way.

FINE (FN): There are no major defects or damage, but some overall, general wear is apparent. It has not been abused and displays well. An item in Fine condition usually has a good resale value, and is generally for collectors who want a well-preserved item at an affordable price.

VERY GOOD (VG): Shows use but no serious damage. However, there may be evidence of fading, worn paint or heavy creases. Although the item is still displayable, the majority of collectors will retain a VG item until they can later upgrade it.

GOOD (G): Obviously worn, with some minor damage in the form of stains, faded colors, chips in paint, missing piece(s), minor dent(s), tear(s), etc. Good items are displayable, but their collectible value is limited.

THE BABY BOOMER CARTOON PHENOMENON

by Hal Lifson

It is really difficult to believe when viewing the cartoon programming on television these days (especially Saturday mornings) that there was once an era in the history of cartoon programming when classics like Rocky and Bullwinkle, The Jetsons, and George of the Jungle were *new* animated programs... surrounded by dozens of others that were equal in their imaginative story telling capabilities.

During the 1950's, when television was still in its infancy, children's programs consisted of both live action hosts with "puppet pals" (Howdy Doody) and television airings of animated theatical shorts along the lines of Bugs Bunny, Tom and Jerry, and Daffy Duck. As the 1950's progressed, and more houses became equipped with television sets, children's entertainment rapidly expanded.

First, the early Fifties brought Crusader Rabbit, produced by Jay Ward and Alex Anderson, which was a cartoon produced expressly for TV airing. While the animation was less detailed than the theatrical shorts, the episodes were well written and the characters Crusader Rabbit and his pal Rags the Tiger were a classic comedy duo.

Advertisers knew they had a captive after-school audience available. Consequently, cartoons were planned and produced in the later 1950's that attempted to capture the attention of the baby boom offspring, both on for television viewing and the in the toy store. Hopefully, kids would then urge their parents to buy licensed merchandise featuring their favorite cartoon pals.

One of the most interesting "made for TV" animated programs was Gumby and Pokey which used the stop-motion animation technique and Play-Doh characters to create a very surreal, yet endearing series of cartoons. While first introduced on the Howdy Doody show, Gumby and his pal Pokey were quickly spun off into their own series, which has survived as a cult favorite into the 1990's. With the exception of the California Raisins, no other stop motion cartoon program has ever captured the interest of American television audiences like Gumby. The Gerry Anderson "Supermarionation" series like Fireball XL-5 and Thunderbirds were also popular, however they were produced in England and have had only intermittent showcases on United States television.

By far, the most significant new cartoon on Fifties TV which set the stage for the astonishing array of wacky and irreverent Sixties cartoons was the 1959 premiere of "Rocky and His Friends" from producers Jay Ward and Bill Scott. Rocky the Flying Squirrel and his dim-witted but lovable pal Bullwinkle moose became overnight cartoon superstars, mainly because of their simultaneous appeal to kids and adults. The Rocky and Bullwinkle episodes were mini-social satires that poked fun at serious issues such as the Cold War, the Soviet Union and prejudice. They also tackled lighthearted social issues that would normally be dealt with on the sitcoms of the era, shows like Ozzie and Harriet, Father Knows Best, and "road" shows like Route 66.

obvious takeoff on The Honeymooners, our friends in Bedrock also dealt with situations very common to live-action sitcom characters and paved the way for a slew of prime time cartoons including Alvin and the Chipmunks, The Bugs Bunny Show, Jonny Quest, Top Cat, and The Jetsons. Even The Bullwinkle Show ran for a year in prime time.

The landscape of children's TV during the 1960's was extremely diverse and inventive. Every year, interesting and, often times, unique characters kept appearing on the TV screen in animated form. Initially, Hanna-Barbera saturated the market with their "wacky animal" characters. Following up on the late Fifties success of Yogi Bear and Huckleberry Hound, the Hanna-Barbera team released instant favorites like Magilla Gorilla, Peter Potamus, Wally Gator, and Touche Turtle.

A company known as Total Television Productions (based in New York) employed the same animation studio that worked on the Bullwinkle and Rocky

The Jay Ward Rocky and Bullwinkle Show format was set up as a variety show, with cliffhanger endings to Rocky episodes sandwiched between other unrelated five-minute cartoon episodes. These episodes spoofed world history (Peabody and Sherman), fairy tales (Fractured Fairy Tales), and assorted children's literature, and even silent film heroes (Dudley DoRight). Principal writers Allan Burns, Chris Hayward and Bill Scott created a fantasy world of clever dialogue, bad puns and memorable characters. All of a sudden it wasn't as important how sophisticated the animation was in TV cartoons, but instead, how accessible the lead characters could become to teenage and even adult audiences.

Things changed very rapidly for television cartoons at the very beginning of the Sixties decade, with the introduction of animation into prime time, with Hanna-Barbera's timeless classic The Flintstones. Presented as an animated sitcom, the characters in The Flintstones family was very accessible and believable to television audiences and was the first thirty-minute animated program on television. An

© Bob Clampett

cartoons, (Gamma Productions in Mexico) to create several new animal cartoons that became immediately successful. Tennesse Tuxedo and His Tales was the first show produced, which was actually one of TV's early educational shows. Tennessee and his walrus pal Chumley would always seek the advice of Mr. Whoopee, who would explain in detail (using the magic 30 Blackboard) how various gadgets operated. The King and Odie, Underdog, Go-Go Gophers, Tooter Turtle and Klondike Kate ("I'll make mincemeat out of that mouse!") all came from the Total Television studios.

Animal characters predominated in the early and mid-1960's--Ideal Toys had the last major "animal hit" of this period, as they were the principal sponsors of the Magilla Gorilla and Peter Potamus Shows produced by Hanna-Barbera. After the prime time ratings smash of Batman in January of 1966, television cartoons rapidly took on the flavor of superheroes and larger than life characters. This trend actually began in 1964 with the impact of the Japanese-produced Astro Boy by NBC and a year later Gigantor, also a Japanese production. Both these characters were of the "Atomic Age" and fought crime in futuristic, often times, gritty circumstances. Even the producer of Felix the Cat cartoons of the early 1960's, Joe Oriolo, jumped on the bandwagon with The Mighty Hercules, a rather cardboard presentation of the mythical hero.

THE 50 BEST ENTRIES WILL WIN WONDERFUL PRIZES AND AN INVITATION TO A SPECIAL CARTOON PARTY AT THE STATION. DON'T FORGET TO FILL OUT THE ENTRY BLANK! CONTEST IS OPEN TO ALL BOYS AND GIRLS WHO HAVE NOT YET REACHED THEIR 13th BIRTHDAY.

MONDAY 6:30 WOODY WOODPECKER	WEDNESDAY 6:30 YOGI BEAR	THURSDAY 6:30 HUCKLEBERRY HOUND

The ALVIN SHOW
PRIME TIME!
NETWORK EXPOSURE!
starring
David Seville and the Chipmunks
ALVIN, SIMON and THEODORE
Licensing Representation:
Nick Draklich CHIPMUNK ENTERPRISES
465 S. Beverly Dr., Bev'ly. Hills, Calif.
© 1962 Ross Bagdasarian

Superhero cartoons had two categories -- serious and obviously silly. In the first category, Hanna-Barbera offered Space Ghost, The Herculoids, Birdman, Might Mighter and The Galaxy Trio while Filmation Productions licensed the classic D.C. Comics characters Spiderman, Aquaman, Batman and friends. Marvel Comics' best known heroes appeared in ridiculously under-animated anthologies (which were popular because of their similarity in tone to the actual Marvel Comics of the period) that featured Iron Man, Sub Mariner, Thor, Captain America and The Incredible Hulk in cartoons that first hit the airwaves in 1966.

On the lighter side, Hanna-Barbera offered Frankenstein Jr. (a minor rip off of Gigantor) and The Impossibles, a rock trio which could instantly transform into the three "super musketeers" Coil Man, Fluid Man and Multi-Man, who would enter into the fray with the battly cry of "Tally Ho!" Terrytoons had the Might Heroes, NBC featured the Super Six on Saturdays in 1966, Underdog continued its successful run (on two different networks in the Sixties and later, in syndication) and, of course, the inimitable Secret Squirrel and Atom Ant (from Hanna-Barbera) and Cool Mc Cool (all on NBC) spoofed both the super spy and comic book hero trends of the mid-1960's.

During the mid-1960's color television became the standard and many houses had a second television set, whose primary function was to entertain the kids. Children's programming was jam-packed into both the early morning and after school hours. Early morning TV in the 1960's was dominated by Captain Kangaroo, a live action CBS show which featured cartoons, but local stations ran reruns of syndicated cartoons and picked up the same format at around 2:30 in the afternoon, which continued until the dinner hour.

TV cartoons became the popular after school baby sitter, and had no other "on screen" competition except for reruns of classic comedy sitcoms like Gilligan's Island, The Munsters and Batman. Ironically, unlike the politically correct Nineties, TV cartoons were very rarely criticized as being unworthy of attracting children's attention when in essence, schoolwork might have been the productive, if less attractive alternative during weekday afternoons. In Los Angeles, for example, several live action hosts were employed by local stations to add some substantive, advice-oriented material to these daily cartoon-a-thons. Hobo Kelly (actually Sally Baker), Mr. Wishbone, Sheriff John, Engineer Bill and Paul Winchell's Winchell Mahoney Time all served as examples of early efforts at creating educational programming for kids.

(Tuesdays ABC-TV Network)

This trend of learning and entertainment came full circle in 1969 with the introduction of Sesame Street, which used animation to augment the learning games and spelling techniques predominant in each weekday's episode. The advent of educationally potent TV forever changed the landscape of children's TV. Even as the superhero trend of 1966 and 1967 gave way to comedy and music driven shows, (like the Archies and The Banana Splits) the classic offbeat cartoons of the Sixties endured, and have continued to permeate the pop culture of the 1990's as they reappear on television channels like Nickelodeon and on home video.

There may never be another era, creatively speaking, such as the one we witnessed in the 1960's, but with constant resurfacing on television today of lost classics we might not need one. We can pour a bowl of Cocoa Puffs and relive the glory days. However, I must admit, once in awhile a "Nineties classic cartoon show" pops in by surprise, such as the "Batman" series on Fox. But don't hold your breath for the new Roger Ramjet...

Top ten classic "crossover" cartoons of the 1960's that can still hold their own against Goof Troop and Chip and Dale's Rescue Rangers:

1. Hoppity Hooper (Jay Ward 1964)
2. Bullwinkle and Rocky (Jay Ward 1956-1963)
3. Cool McCool (King Features 1966)
4. Beatles cartoons (King Features 1966)
5. Linus the Lionhearted (Ed Graham Productions 1965)
6. Beany and Cecil (Bob Clampett 1963)
7. Jetsons (Hanna-Barbera 1962)
8. Marvel Superheroes (Grantray-Lawrence 1966)
9. George of the Jungle (Jay Ward 1967)
10. Batfink (Hal Seeger 1967)

Top ten "hard to find" lost classics of the Sixties animation genre:

1. King Kong (Rankin/Bass 1966)
2. Super President (DePatie-Freleng 1967)
3. Super Six (DePatie-Freleng 1966)
4. Astronut Show (Terrytoons 1963)
5. Magilla Gorilla Show [with opening/closing credits] (Hanna-Barbera 1964)
6. Courageous Cat (Trans Lux 1960)
7. Q.T. Hush (Associate Producers 1963)
8. Rocket Robin Hood (Grantray-Lawrence 1968)
9. Frankenstein Jr. and The Impossibles (Hanna-Barbera 1966)
10. Mighty Hercules (Trans Lux 1963)

TOYS AND TELEVISION

by Paul Matheis

For the youth of that era, the Baby Boom Era was a wonderland of sights, sounds and colors--primarily due to the overwhelming impact of television. Baby boomers became so numerous that they were actually a prime target for advertisers. TV sponsors would go to any length to reach this huge audience of young consumers. Logically, what product would children have wanted most? Toys, of course.

Commercials for a new generation of toys glutted the airwaves after school, on Saturday mornings and seemingly round-the-clock barrages before Christmas. Manufacturers knew where their bread was buttered. In no time, toy packaging carried the banner "As Seen on TV," which almost guaranteed record sales.

Beginning in 1959 and continuing through the early Sixties, **Mattel** sponsored an entire 30-minute weekly kids' show. Over the years, this featured **Harvey** made-for-theater cartoons and TV originals with **Beany and Cecil**. Taking Mattel's lead, **Ideal** joined with animation giant **Hanna-Barbera** to create a $30,000,000 custom-made show solely as a vehicle for their products. The groundbreaking program was called **"The Magilla Gorilla Show,"** debuting in January, 1964. At the time, Ideal vice president **Abe Kent** said, "We're giving the trade wider opportunities for year-round sales and building the foundation for customer acceptance for all our toys and games--even the non-TV items." Ideal president **Lionel Weintraub** added, "We feel that this unprecedented support will be of considerable benefit, not only to ourselves, but to the entire toy industry as well."

"Magilla Gorilla" was an immediate ratings hit. In New York City, it was the highest rated kids' show on the air. Ideal's success with Magilla led to the follow-up **"Peter Potamus Show,"** which premiered in September of 1964. **Milton Bradley** sponsored the popular **"Shenanigans"** show with host **Stubby Kay**. These shows paved the way for later merchandise-based shows such as **"He-Man"** and **"The Transformers"** in the late 1980's.

TV Star Casper

(and company)

Join Famous Doughboy line of Punching Toys

Cash in on Casper, the newest Doughboy punching toy and star of ABC's Number One rated Saturday morning TV show. Casper is pre-sold on TV *every week* in 321 markets, plus as a Harvey Comic Book character he receives millions of additional exposures.

Casper is just one of Doughboy's new 1966 line of fast-selling punching toys. All are made of tough Duraflex vinyl, formulated and manufactured by Doughboy for specific use in Doughboy punching toys.

STARTING
SATURDAY, SEPTEMBER 13TH
FROM 9 TO 9:30 PM ON NBC-TV
CHESTERFIELD CIGARETTES
AND
AMERICAN HOME PRODUCTS
WILL PRESENT

IS A PEGASUS FILM SERIES
FOR TELEVISION ADAPTED
FROM THE NEWSPAPER CARTOON FEATURE

PRODUCED BY DAVID HAFT
EXEC. PRODUCER MICHAEL MESHEKOFF
DIRECTED BY TED POST
BASED ON STORIES AND CHARACTERS CREATED BY MILTON CANIFF
TECH. ADVICE LT. COL. FRANK BALL USAF

STARRING
DEAN FREDERICKS

PRODUCED WITH THE COOPERATION OF
★ THE UNITED STATES AIR FORCE ★

AGENCY: WILLIAM MORRIS

LICENSING FOR MERCHANDISE: TONI MENDEZ, INC.
47 E 61ST. ST. N.Y. 21 TEMPLETON 8-6740

© 1962 Chicago Tribune & New York News Syndicate

THE DICK TRACY 2-WAY WRIST RADIO

Exclusive Dick Tracy Wrist Radio! Millions of junior sleuths across the nation will see TV demonstrations of this electronic transceiver that **really works**—sends and receives up to 700 feet. Dick Tracy's greatest link to crime solution is now the clue to big year-round sales! See this fascinating toy at our showroom, March 5-17.

THE NEW YORK TIMES, SUNDAY, MAY 25, 1958.

NEWS OF TV AND RADIO

'Steve Canyon' Going From Funnies to Film on N. B. C.-TV—Other Items

By VAL ADAMS

STEVE CANYON, the dashing Air Force colonel who is the hero of Milton Caniff's syndicated comic strip, will provide new thrills for television viewers next fall. A weekly half-hour film series titled "Steve Canyon" will be presented on Thursday evenings over the National Broadcasting Company network.

Mr. Caniff, who said he would "sit on top" of the production endeavor, has made available all his comic-strip characters for television use in new situations. on it some more to play Steve. Dana said that would be no problem because his wife owns a beauty shop."

Mr. Fredericks will be following in the footsteps of other young and unknown actors who have gained a certain measure of fame in heroic video roles. It's just possible that Mr. Fredericks will become the Wyatt Earp of the wild blue yonder.

When asked if Steve Canyon's adventures would extend to romance, Mr. Caniff replied they

HA HAHA
HAPPY SALES
PROFITS!..
PEANUTS
PEANUTS
PEANUTS
SNOOPY
AGES 6 TO 12
SNOOPY
SNOOPY
WITH THE FUN GAMES OF THE YEAR
PEANUTS® AND
SNOOPY®!
Charles Schulz' lovable little characters really come to life in these two sparkling Selright games. They're the liveliest, lovingest games you could feature today.
SELRIGHT
SELCHOW & RIGHTER CO. 200 FIFTH AVENUE, NEW YORK 10, N. Y.
Peanuts characters ©1961 United Feature Syndicate, Inc.

*HERE'S BIG LEAGUE MERCHANDISING!

AMERICA'S BEST KNOWN T.V. AND RECORDING FAVORITES, WITH CURRENT PRIME TIME FULL T.V. NETWORK EXPOSURE, AND FULL SATURATION OVER ALL MAJOR RADIO STATIONS IN THE U.S.A.*, ANNOUNCE A NEW MERCHANDISING PROGRAM NOW GETTING UNDERWAY.

THIS IS YOUR CHANCE TO TAKE ADVANTAGE OF THEIR POPULARITY.

REPRESENTATION:

NICK DRAKLICH, CHIPMUNK ENTERPRISES, INC., 465 SOUTH BEVERLY DRIVE, BEVERLY HILLS, CALIFORNIA. TELEPHONE: CRESTVIEW 4-7215

*OVER 16,000,000 RECORDS SOLD

SIMON ■ ALVIN ■ THEODORE ® BAGDASARIAN ENTERPRISES

HANNA-BARBERA'S CARTOON DYNASTY OF THE 1960's

by Hal Lifson (Executive Creative Consultant/H-B Productions)

Beginning in the late 1950's, the Hanna-Barbera reign over television's "cartoon kingdom" began, with the runaway success of Ruff and Reddy (1957) and the next year, The Huckleberry Hound Show. Until this time, cartoons on television were predominantly reruns of theatrical shorts, such as Bugs Bunny and Porky Pig.

Hanna-Barbera's animators created a streamlined look, which might have been viewed as primitive by Disney standards, but the strength of the Hanna-Barbera classic animal cartoons was the cleverness of writing and quality of voice acting, usually performed by cartoon business luminaries Mel Blanc, Daws Butler and Don Messick.

The Huckleberry Hound Show not only introduced a new lead character, but supporting players Pixie and Dixie and Yogi Bear caught on quickly with viewers of all ages. Yogi became so popular in merchandising that in 1960 he got a show of his own, with newcomers Yakky Doodle and Chopper, Snagglepuss, and Hokey Wolf. (A year before, The Quick Draw McGraw Show was launched with Augie Doggie and Doggie Daddy and Snooper and Blabber.)

As early as 1960, Hanna-Barbera had over a dozen characters to promote for licensing and merchandising purposes. There were hundreds of toys, children's games and books with the first wave of Hanna-Barbera classic characters. Reruns of all the cartoons previously mentioned were played throughout the Sixties on local TV stations around the country, and, ironically, seemed to retain their freshness year after year.

All of the classics produced until this point were produced for the first-run syndication market, which generally prevented them from being seen on Saturday mornings (a key showcase for cartoon programming), but instead allowed them to be seen five days a week in the after-school hours.

In 1960, Hanna-Barbera Productions broke new ground with the introduction of television's first prime time cartoon, the instantly classic Flintstones. By combining the best elements of late Fifties sitcoms such as the "Honeymooners" and "The Donna Reed Show,"The Flintstones was actually a modern (for its day) comedy set against a stone-age backdrop. What has made The Flintstones retain its superstar appeal over the last 33 years is its sharp humor and likeable, very believable characters. Classic episodes of the Flintstones when watched today can be viewed as examples of the simplicity of middle class suburban life in the early 1960's. The Flintstones ran on the ABC network for six years, and as early as 1963, was seen on weekdays in syndication as well.

In 1961, Hanna-Barbera introduced its next prime time entry, Top Cat, which was more or less a send- up of the Phil Silvers show, "You'll Never Get Rich." (T.C. sounded a lot like a parody of Silvers' and many of the voices of Top Cat's gang were supplied by actors from the Silvers show.)

The Cats' Meows---"Top Cat," ABC-TV's animated series, had veteran actors Arnold Stang and Maurice (Doberman) Gosfield speaking for two of its leading characters. Stang was the voice of the star performer, Top Cat, and Gosfield spoke for Benny the Ball, Top Cat's pal. "Top Cat" was seen Wednesday nights at 8:30, beginning September 27, 1962.

Although not as big a hit in prime time as The Flintstones, Top Cat has endured over the years as a very popular classic character, with many repeat airings of the original 26 episodes on weekend mornings.

In 1962, two important new H-B classics surfaced. First, another prime time entry, The Jetsons, which of course became another instant classic. Essentially a reworking of the Blondie and Dagwood Bumstead concept (with Mr. Spacely stepping in for Mr. Dithers.) The Jetsons also incorporated America's fascination with the space race and futuristic themes which, beginning in late 1950's, began to influence coffee shop architecture and casual, suburban living room furniture designs. Similar to The Flintstones, The Jetsons offered clever storytelling, imaginative galactic gadgets (many have become standard electronic accessories in the 1980's and 1990's), and likeable characters. Unlike The Flintstones, The Jetsons were not a huge ratings success during its one-season ABC network run.

The second entry in 1962 from Hanna-Barbera was a trio of new animal cartoons presented in a syndicated package called "Hanna-Barbera's New Cartoon Series." It was made up of Touche Turtle, Lippy the Lion and Hardy Har Har, and the irrepressible Wally Gator. These characters quickly joined forces with the other classic animals to create an even stronger army of animated classics. Gold Key Comics, the principal publisher of H-B comic titles in the Sixties, presented a three-issue series called "Band Wagon" which featured Lippy, Wally and Touche. Ironically, these three cartoon characters have not had as much promotion in recent years as some of the earlier characters, but that will most likely change in the Nineties as many of the lost classic cartoon characters will appear in new, updated segments and character merchandise.

In 1964, H-B introduced another breakthrough cartoon with the one-year prime time run of Jonny Quest. This cartoon, certainly one of the studio's most enduring successes, was the first attempt at a

serious, dramatic, live action animation format. Young teen Jonny Quest and his father battled bizarre, prehistoric demons and other-worldly types with the assistance of Race Bannon, Jonny's pal Hadji, and Jonny's dog Bandit. Also the same year, two more 30-minute syndicated, instantly popular cartoon animal shows were introduced by sponsor Ideal Toys: The Magilla Gorilla Show in January of 1964 with supporting characters Mushmouse, Punkin' Puss, and Ricochet Rabbit and The Peter Potamus Show (which picked up the concept of time travel where Jay Ward's Peabody and Sherman left off). The supporting characters on this show were Sneezly and Breezly and Yippee, Yappee, and Yahooey. These two shows ran on weekday afternoons (often back to back), and even had a one-year Saturday morning network run on ABC, with some new episodes added.

Several of Hanna-Barbera's best known characters today had limited runs while in their introductory season, but then were shown for years and years on Saturday mornings with no new episodes produced for many years, if at all. This was the case with The Jetsons, Jonny Quest, Magilla Gorilla, and Top Cat. The likability of these characters allowed kids to enjoy them over and over, while the timeliness of the humor enabled new audiences to discover them.

In 1965, a major new cartoon animal package was assembled, to cash in on the enduring popularity of the James Bond craze and the burgeoning superhero/comic book craze, which would move into full swing in January of 1966 with ABC's live action Batman series. The new Hanna-Barbera series which would run as a one- hour block on NBC Saturday mornings beginning in September 1965 was entitled "The Secret Squirrel-Atom Ant Show" and was broken down as follows: the first half-hour was the Secret Squirrel Show which featured Secret and Morocco Mole as super spies who fought zany villains attempting world domination. Secret's arch nemesis was Yellow Pinkie (an obvious nod to Goldfinger) and the cartoon was a spoof of both the current wave of spy shows and also the Green Hornet who also wore a top coat and fedora and relied heavily on hand-held, high tech, yet non-lethal gadgets to subdue the sinister element. Joining secret squirrel was Squiddly Diddly, who always wanted to leave the aquatic park Bubbleland for a more exciting life in show business, or at least for the big city. Also on this show was Winsome Witch, who tried to do good deeds as one of the few good-hearted witches in the world.

Ideal advertised Magilla's television strangth to promote confidence and even stronger selling power within the industry.

Wait till you see the great new Hanna-Barbera cartoon characters, on NBC-TV this Fall! But don't just wait... grab 'em for your products... don't waste a minute. Complete details are available now.

The second half of this new show was Atom Ant, a tiny, insect, super-crime fighter who could not only leap tall buildings, but could lift them up. Atom Ant quickly became popular because of his diminutive stature as H-B's first "Ant-ie Hero." Joining the tiny titan were Precious Pupp and Granny Sweet and those rollicking rednecks, The Hillbilly Bears, an alternative take on the Clampett family. Especially memorable was the way Paw Rugg used to mumble about everything. The Secret Squirrel and Atom Ant show was previewed in a prime time network special in 1965, which was hosted by Bill Hanna and Joe Barbera, who actually landed on the H-B studio grounds in a helicopter to join Secret Squirrel and Atom Ant in the screening room for a first look at the show. This was the only preview special for a Hanna-Barbera cartoon ever produced.

Other important animated projects of the mid-1960's included two Hanna-Barbera theatrical films. First, "Hey There, It's Yogi Bear" in 1964, was a very charming, clever presentation of the classic Yogi format with excellent big band music conducted by Marty Rich, who worked extensively with Mel Torme and Frank Sinatra. And in 1966, "The Man Called Flintstone" was released as yet another sendup of the spy craze, pinpointing the Bond film "Thunderball" of the year before (1965) for inspiration, with Tania the foreign double agent and the Green Goose plotting to destroy the world with his stolen missile. This film also features some clever set pieces and music, notably Fred and Barney's "Team Mates," a tribute to Hope and Cosby.

Also produced in 1966 was the all but forgotten TV special, "Alice in Wonderland: What's A Nice Kid Like You Doing In A Place Like This?" with Rexall Drugs as the principal sponsor. This one hour, colorful musical featured several celebrity voices including Sammy Davis Jr. as the Chesire Cat (who sings the title song in very hip sequence), Zsa Zsa Gabor as the Queen of Hearts, and Hedda Hopper as the Mad Hatress, the Mad Hatter's mate. This special brings a new animated look to the famed Lewis Carroll classic fantasy, however, the show has only been showcased sporadically on television in recent years. (A home video release is being discussed.) This is certainly the most direct attempt made by Hanna-Barbera in the Sixties to compete directly on Disney turf.

In the fall of 1966, things at Hanna-Barbera would take a drastic turn in yet another new direction, as the previously mentioned superhero craze took hold on the pop culture barometer in the United States.

In the fall of 1966, Hanna-Barbera began a new phase of television programming in an effort to expand the depth and style of its cartoon library. With over 30 wacky animal characters already familiar to television audiences, Hanna-Barbera now attempted to capitalize on the superhero phenomenon by introducing comic book-style characters which resembled what might be found in DC or Marvel comics in the mid-1960's, but were completely original and, consequently, owned outright by H-B Productions.

On the CBS Saturday morning schedule, Fred Silverman (the program director at the time) deemed "Super Saturday" the primary showcase for animated superheroes. Cartoons on the schedule included a new version of Superman from Filmation, Underdog, The Lone Ranger, and from H-B, Space Ghost (a very cool character voiced by Gary Owens), Frankenstein Jr. (update of Japanese 1960's giant robot Gigantor) and the rock group The Impossibles who flipped over their guitars and received commands on a mini television screen from the police commissioner, who would beckon them to save the city from the evil clutches of several different off-the-wall villains. At this point, the rock trio would become Fluid Man, Coil Man and Multi Man.

Space Ghost was certainly the standout new superhero character introduced that season. He wore an all white bodysuit, black hood, and had wrist cuffs that shot laser beams. Space Ghost, who never appeared in his alter identity, had two teenage assistants, Jan and Jace, and a pet monkey, Blip. Space Ghost was set in the future and in outer space, but unlike The Jetsons, this show had a science fiction format wherein Space Ghost would battle intergalactic creatures in his super high-tech spacecraft while at the same time acting as guardian to his two teenage assistants.

Space Ghost had the exact feel of a DC Comics character and was actually created by reknowned comic artist Alex Toth. Space Ghost appeared in his own comic book (from Gold Key) in 1966 and in several issues of the seven-issue series "Hanna-Barbera Super TV Heroes."

Frankenstein Jr. was a clever combination of superhero, classic monster and high-tech robot. Even his voice, performed by Ted Cassidy (Lurch of the Addams Family) had a memorable resonance to it. These 1966 superhero entries were so successful that CBS and NBC followed up the next two seasons with other hard-edged superhero classics like Birdman, The Galaxy Trio, The Herculoids, Mighty Mightor, Young Samson and Goliath, and Shazzan, the super giant genie. Most of the cartoons are currently being aired on the cartoon network and most have not been seen on television in close to twenty years.

After two years of heavy superhero mania, the TV networks decided that Saturday morning programs should have much less violence and, in the fall of 1968, Hanna-Barbera introduced yet another groundbreaking children's show, The Banana Splits Adventure Hour. A musical comedy format, The Banana Splits took off where The Monkees left off, doing comedy sketches, playing music and introducing adventure-oriented segments like the live action Danger Island anthology, The Three Musketeers and Arabian Knights cartoons, and a sci-fi spoof called The Micro Ventures.

Another major H-B entry in the fall of the 1968 season was the Wacky Races, which featured some of the most outrageous race car vehicles seen in animation other than on Tom Slick and Speed Racer of the year before. The featured stars of Wacky Races were Penelope Pitstop (one of Saturday morning's first females heroines) and Dick Dastardly and his sidekick canine, Muttley. Both Penelope and Dick Dastardly later starred in their own shows. The Wacky Races is still very well remembered today as it was rerun consistently throughout the Seventies and Eighties. Kids really responded to the wild gimmickry of the customized race vehicles that were always built around the specific characteristic of its driver.

Following on the heels of these two comedy vehicles of 1968, Hanna-Barbera introduced another classic cartoon show in the 30-minute format (now standard) in the fall of 1969. "Scooby Doo, Where Are You?" was a show built around several teenagers and their crazy adventures with their lovable big canine Scooby Doo (voiced brilliantly by Don Messick) who went on to star in many other series throughout the Seventies and Eighties. Scooby Doo was the last new animal character introduced by Hanna-Barbera in what is known as the "classic period," 1957-1969.

Although Hanna-Barbera continued to produce many other great animated programs in the 1970's, 1980's and now, (a Secret Squirrel update will be on television in the fall of 1993) it is primarily the cartoon shows of the classic era that are best remembered as signature characters and helped to catapault the H-B studios as the number one producer of TV animation in the world.

OVERVIEW OF CLASSIC HANNA-BARBERA CARTOON SERIES 1957-1969

compiled by: Hal Lifson

1957 Ruff and Reddy (syn)
1958 Huckleberry Hound Show (syn)
featured: Pixie & Dixie
Yogi Bear
1959 Quick Draw McGraw Show (syn)
featured: Snooper & Blabber
Augie Doggie & Doggie Daddy
1960 Yogi Bear Show (syn)
featured: Snagglepuss
Loopy De Loop
Hokey Wolf
Yakky Doodle & Chopper
The Flintstones (ABC) 1960-1966 (prime time)
1961 Top Cat (ABC) 1966-1967 (prime time)
1962 H-B "A New Cartoon Series" (syn)
featured: Wally Gator
Lippy the Lion
Hardy Har Har
Touche Turtle & Dum Dum
The Jetsons (ABC) 1962-1963
1964 Jonny Quest (ABC) syn 1964-1965 (prime time)
Magilla Gorilla Show (syn)
featured: Ricochet Rabbit
Mushmouse & Punkin' Puss
Peater Potamus Show (syn)
featured: Yippee, Yappee & Yahooey
Sneezly & Breezly
"Hey There, It's Yogi Bear" (theatrical feature)
1965 Sinbad Jr. (syn)
Atom Ant Show (NBC)
featured: Hillbilly Bears
Precious Pupp
Secret Squirrel (NBC)
featured: Squiddly Diddly
Winsome Witch
1966 Space Ghost and Dino Boy (CBS)
Frankenstein Jr. & The Impossibles (CBS)
Space Kidettes (NBC)
Mighty Mightor (CBS)
"The Man Called Flintstone" (theatrical feature)
"Alice in Wonderland" (ABC TV special)
featured: Sammy Davis Jr.
Zsa Zsa Gabor
1967 Birdman & The Galaxy Trio (NBC)
The Herculoids (CBS)
Shazzan! (CBS)
Abbott & Costello cartoons (syn)
1968 Wacky Races (CBS)
Banana Splits Adventure Hour (CBS)
1969 Scooby Doo, Where Are You? (ABC)
The Catanooga Cats (ABC)
Dastardly & Muttley (CBS)
The Perils of Penelope Pitstop (CBS)

INTERVIEW WITH JOSEPH BARBERA

by Ken Sharp

SHARP: If you could, describe the evolution of animation over the past thirty years and the state of animation today.

BARBERA: *They've been predicting the demise of animation... that's been going on for fifty years. In fact, Friz Freleng, who did Bugs Bunny and all those characters, when he started in the business 50 years ago, his mother said, "Stay out of the business, it isn't going to last," and he always makes a joke about that. He's kind of retired right now. But I know that animation as a business perished theatrically in 1957. There was no more animation or cartoons being made for theater, which was almost a crime. Because I remember we used to preview our "Tom & Jerry's" just like motion pictures, and whenever Tom and Jerry would appear on the screen, the screams, the screams and cheers would go up and people would love them, but then they fell on bad times. The major company, not our studio. And then they closed the studio. Which was actually a great break, since this is the story, the history of animation. What we did, out of desperation, was create a new kind of animation which was cheaper, which was less drawings, which they call "limited animation." So this limited animation, which was of course scorned by real animators, was the only reason the industry survived, because there was no money. To survive you had to create animation with less drawings -- it's that simple. Instead of 26,000 drawings we would have 2,000 drawings, because there was no money! So with this style which we started, we created Ruff and Reddy, then Huckleberry Hound, then Yogi Bear, Quick Draw McGraw, Bob-a-Louie, then came the Flintstones, and right after the Flintstones came Topcat, then Jetsons, Magilla Gorilla, and it just goes on and on. And these vehicles brought the entire industry back. Now our whole studio, which was out of work, gradually was absorbed by our* new *venture, which was very limited in money, very limited in scope, but suddenly took off!*

In fact, our animation was seen in three shows a week -- we were on Mondays with Huckleberry Hound, Wednesdays with Quick Draw McGraw, and we were on Fridays with the Yogi Bear Show, which was a spin-off of Huckleberry Hound, and then we over-lapped that with the Flintstones, and all of this was going on at the same time. It was unbelievable. We were doing, roughly, two hours of animation per week when before we had been doing 45 minutes per year. Forty-five minutes... it was probably less than that. We had been doing eight "Tom & Jerry's," which were five minutes long, and that was... (no wonder I got fired from a bank, I can't even add)... that was 30 to 40 minutes of animation and now we're doing two hours a week. Of course, the happy thing that happened was, when we took our first cartoons to Chicago to show them to an agency who represented the Kellogg Company, we ran the first Yogi Bear and the first "Meeces" -- I don't know if you remember the cat who "hated those meeces to pieces" -- Jinx, Dixie and Pixie? We ran two cartoons for them and they were so funny, the reaction was so big, that they bought the show! And they called it "The Huckleberry Hound Show." And the resurgence of the animation business started at that moment. Now what we didn't know was that we were competing with our own "Tom & Jerry" cartoons which were also being offered to them at the same time. What they did was, they took a chance and bought a totally new package. And how it happened... there were some bars in San Francisco and Gradenia where they would stop serving drinks when that show, Huckleberry Hound, came on. They said no noise, no tinkling of glasses, during the screening of the Huckleberry Hound Show. And throughout the colleges, as you know, they formed Huckleberry Hound Clubs, Yogi Bear Clubs, the phrase "smarter than the average bear," "I'm a non-conformist bear"... And that was soon overtaken by Flintstone and his "Yabba dabba doo," which was a total accident.

For instance, we almost never made "The Flintstones" because it was called "The Flagstones," then they said you can't use "Flagstones" so we switched to "Flintstones." Then we recorded five half-hours and I was totally unhappy with the voices and I went back and got Alan Reed and Mel Blanc. Alan Reed is such a pro and he has such warmth in his voice. One day we were recording, (you see, I directed all the shows for about ten or twelve years, and I wrote a lot of them) and he was looking at the script and I was checking the script in the booth, and he yelled up, "Joe, it says here 'wahoo'." I said yeah, and he said, "Can I say 'yabba dabba doo'?" I said, "Yeah sure, say that." That's how brilliant we are! And that expression has taken off, it's still going. It parlayed, it kept going, and the animation began to thrive, better than ever. We still had people saying "why can't they make 'em with the great animation they used to do?" Because you never had the money! It's so expensive to do animation today. And if someone makes a fully animated feature, you have to be ready to spend between ten and fifteen million dollars. And a lot of people don't want to spend that kind of money. So you have to approach it a different way to survive.

SHARP: I'd like to move into some different areas. That was amazing, you capsulized everything so well. I grew up in the heyday of animation, when it was really starting out, and my love of cartoons still exists today. Why do you feel *those* cartoons, *your* cartoons were the most revered cartoons in the world? The longevity and popularity...

BARBERA: *Well, I wish the <u>networks</u> and the people <u>controlling</u> the animation business would get the message! Why are they still surviving? First of all, they're surviving because they're an expression of our own creativity in this studio, which is the Hanna-Barbera style. The Hanna-Barbera-created characters have personality, and you remember the voices, you remember their particular different style. Like Huckleberry Hound was so laid back, so "Well shucks, things aren't the ways they used to be." He sang "My Darling Clementine" and he was so terribly off-key. So he was one style. Then here comes another brash, con-bear, Yogi. "How do you do, eh, Boo-Boo?" He always had his conscience with him, which was Boo-Boo. Boo-Boo would always say, "The ranger isn't going to like that Yogi" and Yogi would always say, "Can I help it if I'm a non-conformist bear?" So we developed another personality there. Then came the cat who "Hated Meeces to pieces." People didn't know that, but he was a take-off of Brando, used to talk very much like Brando. And he was picking on the cute little meeces all the time, and they had to survive. This was a formula that everyone understood. Then we had the smart-aleck alleycat Topcat, a totally different personality. He was one who survived by his wits, and he had a loyal band of alleycats around him. Do you remember Benny the Bowl and The Brian and ChooChoo, and all these characters? Well that was another style, and these were personalities, you know?*

SHARP: That's something that I think is missing from the other companies and that Hanna-Barbera continues to convey, that personalities are the most important thing.

BARBERA: *Well, to give you an example of how they've survived, we're doing ten features this year for syndication. And when they took one of them, which was called "Yogi's First Christmas"... bears always hibernate, so they never see Christmas. This time he wakes up because there's a big Christmas party going on with his friends up in the lodge and he joins in. Well, that did so well, that our sales people went out to sell it again and they called us and said, "Can you give us ten more." Now no one ever head of anything like that. Who makes a phone call and orders ten two-hour features? I said sure! So what I did was give them 24 titles. And they said no, no, no. We don't want any of those original titles, we want your characters. We want Yogi and we want Topcat and we want Scooby Doo. So this is an incredible feeling, because what the salespeople are selling are characters that we created over 25 years ago. For instance, we're doing "The Good, The Bad and Huckleberry Hound," then we're doing "Topcat and the Beverly Hills Cats." We're doing "The Jetsons meet the Flintstones." Those are all the characters that we've still got rolling.*

SHARP: I wanted to ask you, did you write all the songs for the Jetsons and the Flintstones?

BARBERA: *Yes, Bill and I wrote the themes and the lyrics. We had a marvelous musician, Hoyt Curtain, that did the music, and we've been doing this for over 20 years.*

SHARP: You wrote the music?

BARBERA: *Sure.*

SHARP: And the little jazzy Jetsons?

BARBERA: *Oh yeah! And let me tell you... the Flintstones theme, and the Jetsons theme, those are classics.*

SHARP: You also had 'Stony Curtis' done by Tony Curtis.

BARBERA: *We had Stony Curtis and we had the gentleman who did Stony Carmichael, he was a great musician and pianist. And Cary Granite. The things that helped the show were not only good stories, basic family situation comedy. We did situation comedy before anyone else even thought of it, but the good thing about it is that these stories hold up today. Then we window-dressed the show with the thing that really made it work; for instance, the Stoneway Piano, and the Polarock Camera. The garbage disposal was a prehistoric character sitting under the drain. Cars with no gasoline. You saw the feet, and the feet were pounding away underneath there, when you put the brakes on their feet would smoke up. We had monkeys at stop signs, moving them up and down.*

SHARP: Let's move back to the records you released. Would you ever consider re-releasing any of the classic cartoon records -- The Jetsons, The Flintstones, Yogi Bear, Topcat -- with the original voices?

BARBERA: *Well, under the new set-up that we're setting up right now, musically, it could very well be that we would get out some real nostalgic albums. Or do some MTV using those songs, which would be a very good way to do it. Now we're got four MTV's out now...*

SHARP: What do you mean by that?

BARBERA: *Well, they're MTV's, I don't have the titles, but they're using contemporary songs, like "The Neutron Dance," and we're doing an old one, like the "Monster Mash." And we cut films to it that were from our library, and they work very, very well. There are four new ones.*

SHARP: I wanted to move into the Flintstones, and do a whole thing on them. Tell me how you created the Flintstones. Did you design the characters?

BARBERA: *Yeah. The way that worked was this; we had a sales representative who handled a lot of our meetings with Huckleberry Hound and Quick Draw McGraw, and he came in one day and said, "You know, we've got to get into nighttime. How do we get into prime time with a cartoon? Never been done before. So what we did was to start to create families. And I remember I did at least 80 families, with artists and myself drawing. And you'd have a fat father and a tall mother and a short son and maybe a kid and then it kept going. They'd have sports clothes on, they'd have business clothes on, and nothing worked until one day we had an artist in called Ed Benedict. And we moved into, I don't know exactly how, we said "Let's try stoneage." Those drawings jumped off the page.*

humans and a dog, and the dog became the star. You ask "Well why did he become the Star?" We worked so hard on his personality and we developed this speech where he sort of spoke with "R's". Instead of saying "help" he said "relp". Instead of saying "hello"he said "rero". Well that's something that kids can pick up, and it became his trademark. And you know, people are always ripping things off, but they've never been able to touch Scooby. He's there and he's riding.

And what I did was a lot of the gags, a lot of the visual gags. Like, if you create a "Polarock Camera", that meant there were little birds inside like a woodpecker and he would chisel out the picture inside as you shot it. And I worked on creating all those gags a lot. Ed Benedict created all our model charts for the characters. He had a distinct style, kind of a simplistic style. And what helped everything is that this family became universally accepted because it didn't go into any particular plateau or level, it was a right-down-the-middle-mainstream family. They all had no shoes, they all wore a simple skin. They were identifiable with ever family in the mid, let's say mid-ratio, they weren't rich, they weren't poor, everymid-life American could understand them. And anywhere in the world that they played, Hungary, Sweden, I remember we flew them into Sweden, and we pretaped the Swedish speech. They got off the plane and were gesticulating and talking in Swedish, and everybody understood them.

SHARP: It's so amazing that the Flintstones, their faces, their characters and voices are more recognizable than virtually anything. I mean, you could go to Africa and they'll know.

BARBERA: *They'll know.*

SHARP: They'll know yabba dabba doo, they'll know the Flintstones.

BARBERA: *Absolutely, I tell you. Any country I've been in, when you give them a credit card and they see Hanna-Barbera, they know it's Flintstones, or Yogi Bear. And then, of course, as far as animation went, along came Scooby Doo. That was something I couldn't predict. We created five characters, four*

SHARP: You mentioned Alan Reed a while ago. How close was he to the Fred Flintstone character?

BARBERA: *He sure was Fred Flintstone. He had a warm chuckle, and he was an actor, and he was just distinctively styled to fit that role. I think I just had mentioned previously that I did five shows with two other people. I went back and got Alan and Mel and the minute they got together they were perfect. You know, so many cartoon shows in today's rush of production and tons of animation out there, all of it ground out without any characters out there--you can't pick one character out of them. There's not a lot of styling done on the voices. If you close your eyes and you hear the voices one of them will stand out, one of them will make you smile, and then you're halfway home. But right now, as you listen to voices in cartoon shows, they're all the same. I mean, I could name you the shows, but I won't do that, where you can't tell me who the star is. Because they all talk the same. They have teenagers, and they all talk the same, and the adults all talk the same. The villain has a deep, deep growly voice which comes out of an echo chamber or something, and he sounds the same. And you don't remember any of them. You forget them as soon as you hear them. Now why the network people can't see this I don't know. But you see they have a lot to do with picking the voices. They pick the voices, they pick the characters, they pick the shows. And they really are doing a lot of the... almost producing the shows today. Which fortunately works better for us because that brings back our old character, 'cause that's what people want more than anything else.*

SHARP: I wanted to ask you something I though was really interesting about the Flintstones. If you could discuss how the pop music explosion infiltrated the Flintstones in a way, and I'll give you an example. "The Way-Outs," they were basically a parody of the Beatles.

BARBERA: *Right.*

SHARP: And you had on the impressario Brian Epstein, who was the manager of the Beatles. You gave him, I thing it was either "Eppy Brinestone" or "Brian Eppystone"...

BARBERA: *Yeah, that could have been it...*

SHARP: "The Rolling Boulders"...

BARBERA: *Oh yeah, all of them. The thing about anything you do, is you have to parody, and also ring a bell. To use the name "Rolling Boulders" is about the same as using "Cary Granite"; it rings a bell and it works. And it hangs in all the time. So we did that, every time we could.*

---

and IDEAL brings you the biggest scoop ever—with the newest doll creation...the biggest TV sensation

Coming Feb. 22nd*...the happiest blessed event on nationwide Television...is a blessed event for toy dealers everywhere!

HERE'S WHY! "PEBBLES" makes her debut on one of the top TV network shows . . . a program that's a favorite with both children and adults!

The news of Wilma Flintstone's blessed event will start in January and continue through the birth of "PEBBLES" on February 22nd and straight into March.

Here's an exclusive run-down of the coming Flintstone scripts that publicize Pebbles:

January 25th: Wilma informs Fred that he is about to become a father.

February 1st: Fred's mother arrives to help out in anticipation of the baby.

February 8th: By mistake, Fred hires Foxy Grandma, the bank bandit, to help Wilma.

February 15th: Fred quits his job in order to find a better-paying one in anticipation of the baby's arrival.

*February 22nd: After Fred and Barney go through several "dry runs" in which they rehearse the arrival of the baby and how to get Wilma to the hospital, the baby arrives amid confusion.

March 1st: The Flintstones bring the baby home and Fred "locks horns" with the nurse.

March 8th: Practicing ventriloquism, Barney gives Fred the impression that the baby is talking. Fred thinks he has a genius on his hands.

March 15th: Fred wants to move from the neighborhood to make sure that Pebbles is raised in "upper class" environment...and so on, week after week, in one hilarious episode after another, featuring the Flintstones and Pebbles!

Cash in on the avalanche of sales...when "Pebbles" starts to move!

Have a Cigar!
It's a Girl
at the
FLINTSTONE'S!
VISITORS
ENTRANCE

SHARP: Did you get any reaction from any of the people that you did, such as the Rolling Stones or the Beatles or even Brian Epstein?

BARBERA: *No, we didn't. All we know is that it was just totally accepted, all over the world, and people understood it. Now, when you do that kind of business, when you do that kind of parody, you appeal to a certain group of people. There's another group of people that don't understand it but that are enjoying the show anyway. Now if I show that show in Africa of Japan or wherever where they may not be familiar with the Rolling Stones, or some midwest town, that doesn't bother them, they just accept it because it's a part of the satire.*

SHARP: Would you tell us about the creation of Pebbles Flintstone?

BARBERA: *You see, when you do a network television show, and we were on the air for six years of prime time, which is a record, never been touched by any cartoon show since, you try to do something different every year. Now one of the big things we did one year was we had the birth of a baby. I mean, we were the only ones that ever had a baby born in animation. And I gotta tell you, that show was such a hit. And I remember it was really strange, I remember sitting here in the office and getting a call from New York and the merchandising people saying, "Hey we hear you're going to have a baby born on the show." Nobody ever thought of doing anything like that. I said, "Yeah, it's going to be a boy, we're going to call him Fred Jr.,he's like a chip off the old rock." So they said, "Gee, we have the merchandise people here." They said, "Could you make it a girl?" I said, "Yeah..."*
They said, "You know, they just don't have boy dolls." I never thought of that, but it's true. Girl dolls are what sells. I said, "Gee, I've got to think about that." It took me about three seconds. I said, "Sure, you got it." And then we called her Pebbles, which was a chip off the old rock, too. And that Pebbles doll, that character took off. We had, instead of a clip, we put this little bone in her hair. Those dolls went like crazy! Now, the interesting thing was that the woman, Jean Vanderpyle, that did Wilma, also did the voice of the baby. She did the little baby voice, and she did the mother for it. A very talented woman, I believe, and she's still doing the stuff for us.

PEBBLES 3-D Display Available

The most adorable doll to come along in a million years! PEBBLES . . . the brand new 15″ baby daughter of Fred and Wilma Flintstone. Pebbles' pixie face, and pony tail hair-do with the little pre-historic bone through it, will endear her to everyone. She wears a precious outfit of a sleeveless-smock top and bikini-style panties that let her little tummy peek through. Her jointed arms and legs allow her to be placed in many poses. Pebbles is an enchanting "stone-age" sweetheart sure to be loved by all children.

FREE—Colorful window streamers and newspaper mats!

IDEAL TOY CORPORATION

The Archies

A Chat With The Voice Behind The Archies.

A popular comic book character since the early 1940's, Archie made his cartoon debut on CBS in September of 1968. Aided by pop songs which went on to be number one hits, Archie and gang formed "The Archies" and became one of the top cartoons of the late Sixties, spawning five other half-hour Archie cartoon series from 1968 to 1977. The creative group behind the hit-making songs was The Cufflinks, also known to a select few as Ron Dante!

*The following interview with **Ron Dante** of the **Archies** was conducted by **Ken Sharp**.*

Ken: So the **Cufflinks** was just you?

*Ron: Yeah. **Ropper Holms** did some of the arangements for the Cufflinks and I think, in the first album, there was one cut that came from nowhere. Somebody else was singing and the producers decided to just put another song on and they never called me so it was like, nine cuts by me and one cut by the ghost group that I never heard of.*

Tell me how you got involved with the **Archies** and what was the first song you cut. Was it **"Truckdriver"?**

*This was a long time ago. I think it was "Truckdriver",I'm pretty sure. I had to audition for "The Archies". I had to go in and face **Jeff Berry**, **Don Kirschner**, and the engineer and the arranger. Jeff Berry got me in the studio and said, "Now try this song. I want you to try it hard." I sang hard. "Now soft." I sang it soft. "Now Higher." I sang it higher. I did everything. I needed that job at that time. Jeff and Donny, I could see them talking about me and I couldn't hear what they were saying through the glass, but I could hear them talking. Talking about me saying, "Well, could he do it, could he not do it." And I remember thinking, "If I get this, this is going to be great, but I hope I get it because, you know, I had to audition. Finally, Jeff came out to the studio and said, "Yeah, we're going to do it very breathy." "Very breathy," I said, "Fine, anything you say." So that was the way I got involved.*

But did you know at the time it was for a cartoon group?

Yes, I knew it was from a cartoon series. I knew it was for Archie, one of my favorite characters from the comic books. I grew up on Archie and they said it would be on every Saturday morning. It was a children's thing and I really never anticipated pop hit records out of it. I thought it was just going to be the music; just another gig.

But the songs caught on and even made the top ten on the music charts?

*Yes, **"Sugar Sugar"** was number one, the Cufflinks were number five, and people still came up to me and said, "Well, what are you doing with your life?" And I said (laugh), "I have the number one and five record!" It was sort of tough to get through that time because I was a "ghost." But I was a very famous ghost at the time.*

Archies Cut-Out Paper Dolls
(Whitman, 1969)

Did you ever do any demos for songs that became real big hits?

Well, I did one. The songwriters came to me and said, "We want you to sing this song for ***Connie Francis.*** *" I said, "Fine, when do we go into the studio," and they said, "No, no, we don't go into the studio. We pay you $20.00 to learn the song and you go down the street and sing it live for Connie Francis." So, (laugh) I learned a song called* ***"I'm Going To Be Warm This Winter"****, one of her many hits. I went down and there was another girl named* ***Ellie Greenwich*** *who had to sing another song live. So there was a live demo, and I got $20.00 cash to sing for Connie Francis.*

What was a typical Archies recording session like?

Well, in the beginning, the first sessions were all live musicians. Jeff Berry producing all of it and writing a ton of it. He wrote with ***Andy Kim,*** *(who) did a lot of his writing with him and* ***Bobby Bloom*** *and there were about five musicians. Jeff had a definite ideal of what the sound should be: it should be the cartoon characters playing the guitar, base, drums and tambourines, and that was what was going to be on the record. If you listened, very rarely was there anything else. No trumpets, no violins, no extra things, no organ, and it's just a five piece rhythm section and that's what he made all his records from.*

You know, it's very interesting, a lot of people don't really talk about it, but the later Archie stuff was very socially conscious. A lot of people don't realize it, they think it was a lot of meaningless, fun songs, which a lot of it was. There is nothing wrong with that, but a lot of the songs were actually socially conscious. Was that an interesting avenue to pursue?

I liked that. I thought it was something I had to do, consciously, because we were reahing millions of children every week and every Saturday morning. Those kids are very impressionable and I think they wanted songs about air pollution, songs about violence and world peace. There was one song we did called ***'Summer, Pray for Peace."*** *At least we got to the young minds and we tried to make something of it and put it together in songs.*

What do you remember in the session in the recording of **"Sugar Sugar"**. What a smash it still is today!

The only thing I remember about it was the vocal. We paid a lot of attention to very small things in the song, like the breathy sound. When I took a breath to breathe, I had to make it pronounced. So you could hear me breathe at some points. It happened so quickly, and all of a sudden it was number one.

Now, what's the great **Ed Sullivan** story?

My mother called me up one night a few months later and said, "You are going to be on the Ed Sullivan Show" and I said, "But I can't be on Ed Sullivan (laugh) I'm here at home watching it with you." and she said no,no the Archies are going to be on with Ed Sullivan tonight." And sure enough Ed Sullivan comes out and says, "Right here, The Archies!" and they played the cartoon of The Archies and there was my voice right there on the Ed Sullivan show. So my dream was to be on the Ed Sullivan show, but at least I got my voice on it.

I'm going to ask you about some tracks. Do you recall **"Jingle Jangle"**? Was that **Tony Wines** singing with you?

"JingleJangle"was done in the wrong key. It was one of those things they did before I arrived, and Jeff, I don't know what happened to him that day. He did the track in the wrong key and he said, "I want you to sing it no matter what." I said, "But it's up there in my falsetto. I'll sound like a girl." And he said, "That's fine." So I ended up singing the whole track of "JingleJangle"and putting multi track in my voice three or four times and I did all the lead stuff on "Jingle Jangle" and everybody said, "That must be Tony," but it was really me. The Archies must have cut 200, 300 songs in three years and they all kind of blended in together. They were done in such rushes. Each time was a concentrated effort.

The next song I wanted to mention, which you wrote, is a great song, **"Everythingis Alright".**

Yeah, I wrote that for a friend of mine who was teaching at a school for retarded people and I wrote that for the people there. I went there and I would sing that to them on Saturday mornings, not realizing I could cut that. So I finally, at one session, played it for Jeff and he said, "Yeah , let's cut this, it's fine." So we did end up cutting it but I really did write it as a therapeutic song.

In your eyes, what do you think made up that kind of "bubble gum" sound? Do you think people wanted something that was a bit more innocent at the time, because it was a pretty turbulent period?

Times were tough then. It was the years of the war, the political unrest, and people were scared. I think this kind of fairly innocent music was something they could relate to. It was fun to dance to, it was associated with an American icon--the Archie Comics, which was All American. People had grown up with it , so I think the image of Archie and the music all together gave people some kind of serenity, they felt a little better about the world. In fact, the hard rock of the time didn't become the biggest record of the year--the biggest record of the year was "Sugar Sugar".

Did you save any Archie memorabilia?

I have some of it. I have a lunch box and I have records and things. I have some of the memorabilia, but most of it is gone.

ALVIN & THE CHIPMUNKS

Already popular record album stars on Liberty Records (1959-1960), Alvin, Theodore and Simon found new success in their own cartoon series which premiered on CBS in October of 1961. The half-hour cartoon show featured two Alvin cartoons with sing-along segments and one episode of the crazy inventor, Clyde Crashcup.

1

3

4

1) ALVIN AND THE CHIPMUNKS SOAKY BUBBLE BATH CONTAINERS (1963) Set of three 10" plastic figure bubble bath containers of Alvin, Simon & Theodore. EACH: **$10-20**

2

2) ALVIN AND THE CHIPMUNKS "SOAKY" RECORD (Colgate-Palmolive/Soaky 1963) 6"x6" illustrated thin cardboard record contains songs and greetings from the Chipmunks and features a full color illustration of the Chipmunks and David Seville. This record was offered as a mail order premium from a Colgate-Palmolive series of Chipmunk bubble bath containers. **$25-50**

3) ALVIN AND THE CHIPMUNKS CHRISTMAS STOCKING (1963) 14" long plastic vinyl hang-up Christmas stocking with colorful illustration of Alvin, Simon and Theodore each playing an instrument and a word balloon overhead saying, "Merry Christmas from Theodore, Simon and Alvin." **$25-35**

4) ALVIN FOR PRESIDENT 45 rpm RECORD (Liberty 1960) 7"x7"paper sleeve contains 45 rpm record with the song "Alvin for President" by David Seville and the Chipmunks. **$10-15**

5

5) ALVIN "ACORN HUNT" BOARD GAME (Hasbro 1960) **$35-40**

6

6) **THE THREE CHIPMUNKS "BIG RECORD" BOARD GAME** (Hasbro 1960) **$35-40**

7) **ALVIN HAND PUPPET** (Knickerbocker 1962) 10"tall puppet with cloth body and molded vinylhead. **$25-35**

8

8) **ALVIN LUNCHBOX WITH THERMOS** (1963) Green vinyl box: **$250-350** Steel thermos: **$50-75**

9

9) **ALVIN MUG** (1959) Heavy glass mug features colorful scene of early Alvin walking down a country lane. **$15-30**

10

10) **THE THREE CHIPMUNKS PAINT AND CRAYON SET** (Hasbro 1959) 10"x12" box contains pre-numbered sketches, eight watercolor paint tablets and brush. **$25-50**

11

13

11) **ALVIN RECORD ALBUM "THE ALVIN SHOW"** (Liberty 1962) 33-1/3 rpm record features theme song. **$15-25**

12) **ALVIN RECORD ALBUM "SING ALONG WITH THE CHIPMUNKS"** (Liberty 1962) Red vinyl. Scarce. **$15-20**

13) **ALVIN RECORD LP "CHIPMUNK SONGBOOK"** (Liberty 1962) 7"x7"cardboard sleeve contains 45 rpm record with four songs. **$10-15**

14

14) ALVIN STORE DISPLAY BOX (Kenner 1962) Display box for Kenner's "Give-a-Show" projector accessory slides. Large 15"x15"x12" box with many other cartoon characters as well. **$25-50**

15

16

15) ALVIN STORE DISPLAY FIGURE (1959) Beautiful 24" tall detailed figure of Alvin with removable metal straw hat that fits on head or either hand. Used in conjunction with Alvin's debut record on Liberty. **$100-200**

16) ALVIN AND THE CHIPMUNKS TATTOO WRAPPER (Fleer Corp. 1966) 1.5"x3.5" tattoo wrapper features portrait of Alvin on front of wrapper. **$25-50**

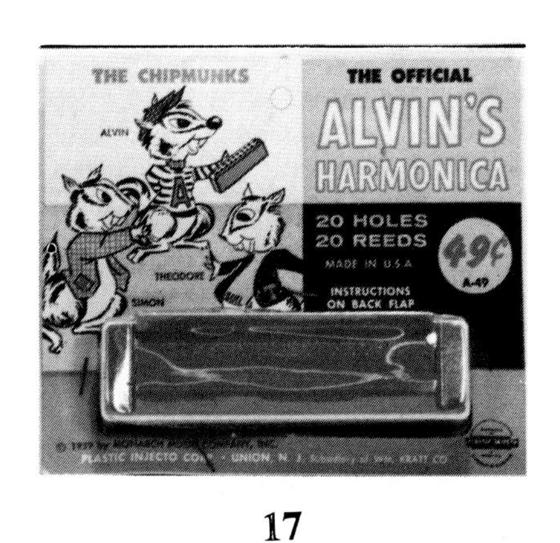

17

18

17) THE OFFICIAL ALVIN'S HARMONICA (Plastic Inject Corp. 1959) 5"x5.5" illustrated display card holds 4" plastic blue/white harmonica (20 holes/20 reeds). Card features early illustrations of Alvin, Theodore and Simon. A very early Chipmunks item. **$25-35**

18) CHIPMUNK CHALK (Rel 1962) 4"x4.5"x2" box contains color chalk. **$15-25**

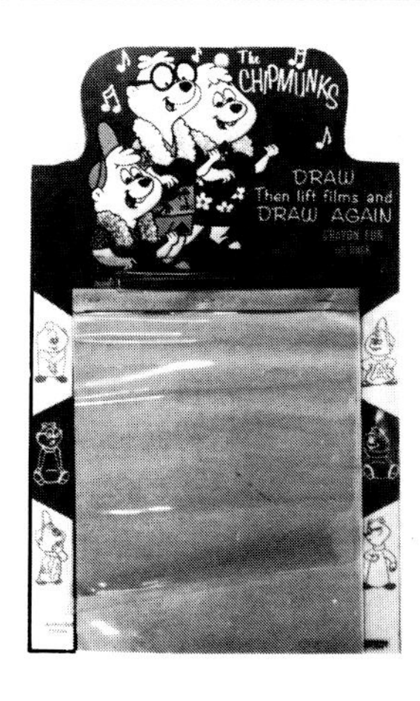

19

19) THE CHIPMUNKS MAGIC DRAWING SLATE (Saalfield 1962) 8"x12" illustrated cardboard display card holds lift-up erasable film sheet and comes with wood stylus. **$15-25**

20) ALVIN AND THE CHIPMUNKS GREETING CARDS WITH RECORD (Buzza 1963) Colorfully illustrated 5"x5" cards contain small record inside with extended message/greeting from the Chipmunks. Cards include; get well, happy birthday, etc. Label of record is also illustrated with the same scene as the cover of greeting card. Card made of hard glossy heavy cardboard. **EACH: $5-8**

BEANY & CECIL

Animator Bob Clampett originally created Beany and Cecil back in 1949 for a live action puppet show entitled "Time for Beany" which ran five days a week for over five years. In 1961, United Artists and ABC collaborated with Clampett to produce an animated series. Mattel Toy Company, knowing the success of the early puppet show, backed the project and secured many exclusive merchandising rights. In fact, when the cartoon first premiered, it was titled "Matty's Funday Funnies," then "Matty's Funnies with Beany and Cecil," to finally "Beany and Cecil" (Matty Mattel was Mattel's official mascot). Beany and Cecil premiered on ABC in January 1962.

1

1) "BEANY & CECIL AND THEIR PALS" RECORD PLAYER (Vanity Fair 1961) 6"x11"x13"fold-open child's record player made of thick cardboard with litho paper design, adaptable for 45 or 78 rpm records. **$100-200**

2

2) BEANY & CECIL BATH MITT (Roclar 1962) 8"x4" window display box contains green terrycloth childs mitt in the likeness of Cecil. Also included is an assortment of small cakes of soap in the shape of Beany and Cecil. **$35-50**

3

3) BEANY & CECIL "BEANIE COPTER" HAT (Mattel 1962) Red plastic cap that launches propeller. **$50-75**

4

4) BEANY & CECIL BUBBLE BATH (Roclar 1961) 12"x6"x3"window display box contains 24 illustrated packets of scented bubble bath in assorted colors. **$35-50**

5

5) BEANY & CECIL CARTOON KIT (Colorforms 1962) 8"x12"box contains die-cut thin vinyl stick-on character pieces and accessory pieces, plus background scene for placement of characters, and fold-out instruction booklet. **$50-75**

7

6

6) BEANY & CECIL "CECIL" HALLOWEEN COSTUME (Ben Cooper 1961) 10"x12"box contains mask and one-piece fabric bodysuit with illustration of Cecil on front. **$50-75**

7) BEANY & CECIL "CECIL" BUBBLE BATH CONTAINER (Purex 1962) 10" plastic soap container. **$20-30**

8

9

8) BEANY COLORING BOOK (Artcraft 1963) 8"x11",100-pages. **$20-30**

9) BEANY & CECIL COLORING BOOK (Whitman 1953) Early Clampett 8"x11",100-pages. **$30-45**

10

10) BEANY & CECIL TARGET BALL SET (1962) 14"x14" colorful tin litho tray with three wooden sides. There are several circular indentations into which balls fall to score points. Set comes with five colored wooden balls in 15"x15"x3" box. **$50-75**

11

11) BEANY & CECIL 3-D MOSAICS (1961) 8"x4"box contains three bags colored crushed stone, a 5x7" pre-numbered sketch of Cecil and plastic frame. **$50-60**

12

12) BEANY & CECIL "LEAKIN' LENA" BOAT (Irwin 1962) Extremely colorful 15"x18"display stand holds large plastic 15"x15"boat that is propeller-driven by rubber band, allows it to float and roll across the water. **$100-150**

19

19) CECIL AND HIS DISGUISE KIT (Mattel 1961) 10"x16" window display box contains 24" green plush bendable figure of Cecil plus 18 different disguise costumes which includes hats, wigs, glasses, capes, beards, etc. **$50-75**

20 21

20) TALK TO CECIL GAME (Mattel 1961) 14"x15"window display box contains 12" green plush talking Cecil handpuppet, a 17-piece interlocking Cecil puzzle playing board, 14 playing cards, four Beanie figure playing pieces, card rack and die. Object of the game is to be the first player to rescue Beany. **$75-100**

22

24

21) CECIL GE-TAR (Mattel 1961) 18" tall hard plastic guitar with concealed music box which is crank-wound. Guitar face has full color die-cut lithographed paper label of Cecil with Beany and characters in boat. When crank is turned, Cecil's eyes move and his tongue wags to the beat of the melody produced. **$20-30**

22) CECIL-IN-THE-MUSIC-BOX (Mattel 1961) 5"x5"x6" litho metal box with crank handle produces a tune before cloth figure of Cecil pops out of box. **$50-75**

23) TALKING CECIL DOLL (Mattel 1961) 18" tall doll made of green plush with blue fabric tail fins and eyelashes with free moving eyes, features a voice-activated pull string which produces eleven different phrases. **$75-100**

24) DISHONEST JOHN FRAME TRAY PUZZLE (Playskool 1961) All wood inlay jigsaw puzzle in a 12"x13" frame tray. **$20-35**

26

25) DISHONEST JOHN "JUMPIN' DJ" GAME (Mattel 1961) 15"x10"boxed game includes a flexible figure of DJ that can be compressed to a smaller size and, over the span of a couple minutes, releases to original size, causing the figure to spring up and "jump". Also comes with illustrated tin litho scoreboard. **$20-40**

26) DISHONEST JOHN TALKING HAND PUPPET (Mattel 1961) 18" tall cloth body puppet with molded vinyl plastic head and hands has a pull string in back which activates a concealed voice box producing a variety of phrases. **$50-100**

13

13) BEANY & CECIL MATCH-IT PUZZLE TILE GAME (Mattel 1961) Large 18"x12"tray puzzle-style game. Each square has Beany and Cecil on all four sides and the object is to match each side up with its identical matching piece. **$15-25**

16

17

16) BEANY & CECIL VINYL CARRYING CASE SET (1961) Pair of red vinyl cases, one a small rectangular 5"x3-1/2"x3-1/2" case with all four sides illustrated with characters. The second case is a large circular shape and is 8" in diameter with llustration on one side only. PAIR: **$50-75**

17) BEANY & CECIL WASH CLOTH AND BUBBLE BATH (Roclar 1961) 5"x16" package with display card contains four-color cloth wash cloth depicting Beany and Cecil and their Leakin' Lena ship. Also eight illustrated packets of scented bubble bath in assorted colors. **$25-40**

14 15

18

14) BEANY & CECIL SOAP ON A ROPE (Roclar 1961) 6"x6"x3"box contains 5" tall green soap molded in the image of Cecil with a green/white fabric rope necklace extending from his head. Also included is an assortment of smaller soap figures in the shape of sea shells, sea horses and fish. **$25-50**

15) BEANY DOLL (Mattel 1962) 18" stuffed cloth doll with molded plastic vinyl head, hands and shoes has a small plastic propeller on the back of his cap which actually turns. This doll was also made as a talking doll which had a voice activated pull string which, when pulled, produced eleven different phrases. DOLL: **$50-100** TALKING BEANY DOLL: **$100-150**

18) CECIL PLUSH DOLLS (Mattel 1961) Two different non-talking dolls were produced by Mattel and made of green plush with blue fabric eyelashes and tail fins with free moving eyes:

A)15" tall, coiled position to stand by himself **$20-30**

B)24" long, slender, can be shaped in various poses **$30-65**

BOZO THE CLOWN

Originally a Capital Records and comic strip favorite dating back to the late 1940's, Bozo was revamped by creator Larry Harmon in 1959 for a half hour cartoon show which revolved around the circus. Bozo also had sidekicks Butch the Circus Boy and his dog, Elvis.

1

1) **BOZO BUTTON** (circa 1960-61) Colorful 3.5" diameter metal button reads "Bozo the Clown." **$8-12**

2

2) **BOZO THE CLOWN CIRCUS GAME** (Transogram 1960) 8"x16"box. **$20-30**

3

3) **BOZO THE CLOWN "KID KLEANER" BUBBLE BATH** (Riley 1960) 8"x6" illustrated box contains powder bubble bath. Color graphics on all six sides of box. **$40-50**

4

4) **BOZO THE CLOWN "KING OF THE RING" BOOK** (Tell-a-Tale 1960) 5"x6" book with 20+ page story accompanied by full color story art on each page. **$6-10**

5

5) **BOZO CARTOON KIT** (Colorforms 1960) 10"x12"box contains stick-on character figures and background board. **$35-45**

6) **BOZO LUNCHBOX WITH THERMOS** (1962) Metal dome-top box: **$100-200** Thermos: **$20-40**

7

7) BOZO NUMBERED PENCIL COLORING SET (Hasbro 1968) 9"x10"box contains twelve pre-numbered sketches, ten color pencils and sharpener. **$20-25**

8

8) BOZO RECORD ALBUM SETS (Capitol 1950) 12"x10" cardboard booklet and 20-page illustrated story contains two 78 rpm records which contain matching story. EACH: **$15-25**

11

11) BOZO TALKING TRICYCLE (Stelber 1961) All-metal child's tricycle features a hard plastic battery-powered bust of Bozo mounted in front which, when activated by button, produces eight different phrases. Rear hubcaps have embossed star inside a circle and the handlebars have plastic red, white and blue streamers. The front of bike has red/white decal of Bozo name directly below bust. **$300-400**

9 10

12

12) BOZO TV-LOTTO GAME (Ideal 1960) 14"x10"box contains illustrated square tiles and playing boards of Bozo and his gang. **$20-25**

9) BOZO BUBBLE BATH CONTAINER (1963) 10"plastic soap container with hard plastic removable head. **$15-25**

10) BOZO STORE DISPLAY STATUE (Capitol Records 1950) Large two foot tall well-sculpted colorful ceramic figure of Bozo used by large retail stores to promote the Bozo records. **$75-125**

CEREAL CHARACTERS

The late Fifties and Sixties saw a tremendous rise in cartoon characters on the boxes of children's pre-sweetened cereal.

Some characters, such as Twinkles the Elephant (1960-63) began as actual cartoon characters and were licensed out. (Twinkles was originallyfeatured on the King Leonardo and His Short Subjects cartoon show.) Conversely, some cereal box characters such as Linus the Lionhearted, (Post Crispy Critters 1962-72) grewin popularity to the extent of having their own Saturday morning cartoon shows and a variety of toys produced in their honor.

Bullwinkle creator Jay Ward animated and produced the popular Cap'n Crunch, Quisp and Quake cartoon commercials of the Sixties and added a whole new dimension to cereal commercials, their characters and their demand.

1

4

1) **CAP'N CRUNCH BANKS** (Quaker 1966) Plastic 7" color figures of Cap'n Crunch and Jean Lefoot which were available as a mail order premium. EACH: **$15-25**

4) **CAP'N CRUNCH HAND PUPPETS** (Quaker 1964) 8" thin plastic puppet which came free inside specially marked packages of Cap'n Crunch. A series of six crew members were made: Cap'n Crunch, Seadog, Brunhilde, Alfie, Carlyle, and Dave. Our photo shows Cap'n Crunch. EACH: **$5-10**

2

3

5

2) **CAP'N CRUNCH BUTTON** (Quaker 1965) 3" diameter button with color illustration of Cap'n Crunch's head against a white background. This was part of a membership kit issued by Quaker in the mid-Sixties. **$10-12**

3) **CAP'N CRUNCH CEREAL BOX** (Quaker 1963) "Stays crunchy...even in milk" was Quaker's slogan for introducing its Cap'n Crunch cereal into the market when it first debuted in 1963. By 1965, just two years later, it was the second largest selling pre-sweetened cereal in the country. The first issue box advertised one of nine different pirate rings inside. **$75-150**

5) **CAP'N CRUNCH KALEIDOSCOPE** (Quaker 1964-65) Colorful 7" long cardboard kaleidoscope with illustration of all six characters (Cap'n Crunch, Seadog, Brunhilde, Alfie, Carlyle, and Dave) on both sides. This was a mail order premium. **$20-35**

6

6) CAP'N CRUNCH "OATH OF ALLEGIANCE" PLAQUE (Quaker 1964) 8"x10"plaque made on high-grade document paper with color illustration of Cap'n Crunch surrounded by gold leaf cluster on top and affixed seal of acceptance on bottom. The oath requires the bearer of the plaque to "be good and eat plenty of Cap'n Crunch." This is part of an early membership kit. **$10-20**

7

7) CAP'N CRUNCH SEA CYCLE (Quaker 1960's) Plastic snap-together assembly kit of working sea-cycle that is rubber-band powered and skims across water. There is also a figure of Cap'n Crunch and Seadog which sits on top and rides the cycle. Kit is 6"x5"high when assembled and comes with color stick-on decals of Cap'n Crunch and Seadog. This was a mail order premium. **$25-35**

8 9

8) CAP'N CRUNCH SHIP SHAKE (Quaker 1968) Clear plastic 6"tall tumbler with red lid features color illustration of Cap'n Crunch on side. This was a mail order premium. **$15-25**

9) CAP'N CRUNCH FIGURE RING (Quaker 1963) One of the very first premiums to be offered in Cap'n Crunch cereal was a series of nine different rings, all featuring a sea-pirate related theme (Ship-in-a-Bottle Ring, 1863 Gold Coin Ring, Treasure Chest Ring, etc.). The figure ring of Cap'n Crunch is 1" tall and stands on top of ring. **$15-20**

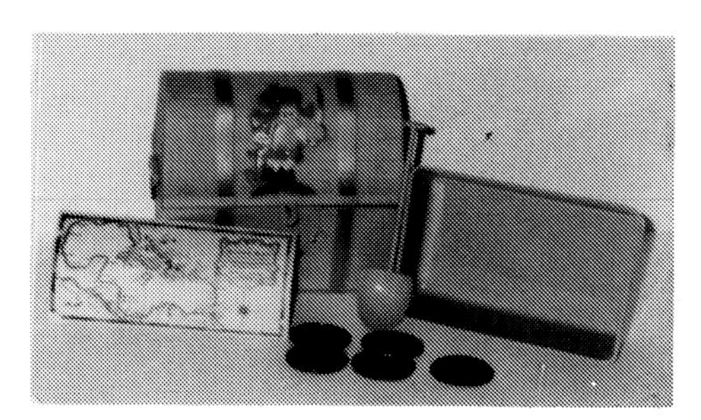

10

10) CAP'N CRUNCH TREASURE CHEST BANK AND CEREAL BOWL SET (Quaker 1960's) 6"x5"x4"deep tan plastic treasure chest bank opens up to reveal a tray-like cereal bowl and shovel-shaped spoon. Also included are several plastic gold pirate coins and a paper treasure map. There is an embossed color figure of Cap'n Crunch on the top of the chest and a working lock and key holds latch closed. This was a mail order premium. **$35-50**

11

11) THE CHEERIOS KID AND BULLWINKLE CUP AND PLATE SET (General Mills 1965) 8" diameter white plastic plate has full color illustration of Bullwinkle on diving board and Cheerios Kid standing beside pool. Cup is white plastic and features black and white illustrations of Bullwinkle, the Cheerios Kid and Sue all running. PLATE: **$10-15** CUP: **$12-18**

LINUS THE LIONHEARTED

From Post cereal mascot for Crispy Critters to television star, Linus the Lionhearted premiered as a half-hour cartoon on September 26, 1964. Supporting characters were other Post cereal mascots including Sugar Bear (Sugar Crisp), lovable Truly the Postman (Alpha Bits) and So Hi (Rice Krinkles).

CRISPY CRITTERS CEREAL BOX (Post 1964) "The one and only cereal that comes in shapes of animals." This slogan, when shown on the TV commercials, started a stampeding herd of animals across the television screen, trampling poor Linus the Lionhearted in the process. Introduced into the market in 1962, our photo shows a third year box. **$50-75**

12

12) LINUS THE LIONHEARTED CRITTER CARDS (Post Cereals 1965) Deck of playing cards with illustrations of Linus the Lionhearted and other jungle animals comes in 3"x4" illustrated protective box. These cards were offered as a mail order premium from specially marked boxes of Post Crispy Critter Cereal. **$20-30**

13

14

13) LINUS THE LIONHEARTED FIGURE BANK (TransOgram 1965) Yellow 10" plastic bank. **$20-25**

14) LINUS THE LIONHEARTED FRAME TRAY PUZZLE (Whitman 1965) Inlay jigsaw puzzle in a 11"x14" frame tray. **$15-30**

15

15) LINUS THE LIONHEARTED FUN BOOK PREMIUM (Post 1964) 5"x7"booklet contains 30-pages of black and white stories to color, puzzles and games. Set also contains a small Colorforms set which consists of die-cut thin vinyl stick-on character pieces of Linus the Lionhearted and three other jungle animals and circus accessories plus a background circus scene for placement of characters. Originally came in 5"x7" illustrated brown mailing envelope. This was offered as a mail order premium from specially marked boxes of Post Crispy Critters. **$25-50**

16

16) LINUS THE LIONHEARTED LUNCHBOX WITH THERMOS (1965) Green vinyl box: **$200-300** Metal thermos: **$50-75**

17

17) LINUS THE LIONHEARTED RECORD ALBUM (Post 1965) Linus and his other Post cereal friends. **$15-30**

18) LINUS THE LIONHEARTED STUFFED DOLL (Post 1965) Cereal mail order premium of 12" stuffed Linus. **$25-50**

19

20

19) LINUS THE LIONHEARTED TALKING DOLL (Mattel 1965) 18" brown stuffed plush doll with molded vinyl plastic face and hands has a voice-activated pull string which, when pulled, produces eleven different phrases. **$100-125**

20) LINUS THE LIONHEARTED TALKING HAND PUPPET (Mattel 1965) 12" plush puppet with hard plastic face has pull string that activates concealed voicebox, producing a variety of phrases. Window display box is 7"x12"x5"deep. **$40-50**

21

21) LINUS THE LIONHEARTED "UPROARIOUS" GAME (Transogram 1965) 10"x19"box also features other Post cereal characters including **Honey Bear, Truly the Mailman, Granny** and three others. Game includes four stand-up figures of Linus, vinylGranny bag and playing cards. **$75-100**

22

23

22) QUAKE CEREAL BOX (Quaker 1965) Quake was introduced in 1965 along with his adversary Quisp. Quake, the earth-digging miner mascot, was created by Jay Ward of Bullwinkle cartoon fame. Quake, a corn based cereal, didn't have the taste kids wanted, and after a grand, pre-conceived, national popularity contest between Quisp and Quake cereals, in which kids were to vote for their favorite character, Quake lost. The cereal was discontinued in 1975. **$200-350**

23) QUAKE DOLL (Quaker Oats 1965) 10" tall stuffed Quake doll. **$40-80**

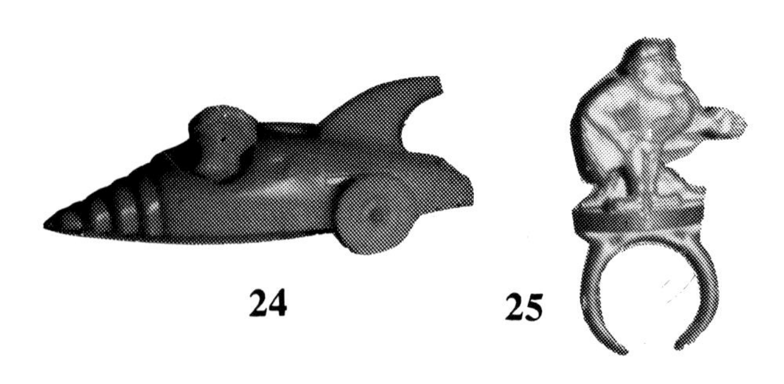

24 25

24) QUAKE EARTH-DIGGER CAR (Quaker 1965) Plastic 2.5" car with drill-bit front and fin in back holds a small detailed figure of Quake in his cowboy-style hat. There is a launcher that comes with the car and shoots the car forward. This premium came free in boxes of Quake. **$50-100**

25) QUAKE FIGURE RING (Quaker 1966) Small detailed plastic figure of Quake sits on top of plastic ring band and came free inside boxes of Quake cereal. **$50-100**

26) QUAKE MINER'S HELMET (Quaker 1966) Yellow plastic miner's helmet with battery-powered light on front and embossed words "Quake" on sides. This was a mail order premium available through Quaker. **$100-150**

27

28

27) QUANGAROO FREE-WHEELER (Quaker 1964) 2" plastic snap-together figure on wheels. **$25-50**

28) QUISP CEREAL BOX (Quaker 1965) Quisp debuted in 1965 and was an instant success, partially due to the witty TV commercials that flooded Saturday morning programs. The little alien mascot, Quisp, was designed by Bullwinkle creator Jay Ward, who was also behind those clever, animated commercials. Quisp was in production for twenty years and continues to be the most widely requested cereal of all time. Our photo shows a 1969 box which features the Quisp Gyro Unicycle premium. **$100-200**

29) QUISP CERAMIC BANK (Quaker 1960's) Colorful 6" ceramic figure of Quisp on base. **$50-100** (Note: Reproductions of this bank are currently being produced and are sometimes passed off as originals. The reproductions are quite good and buyers should look closely for mold flaws and edge lines.)

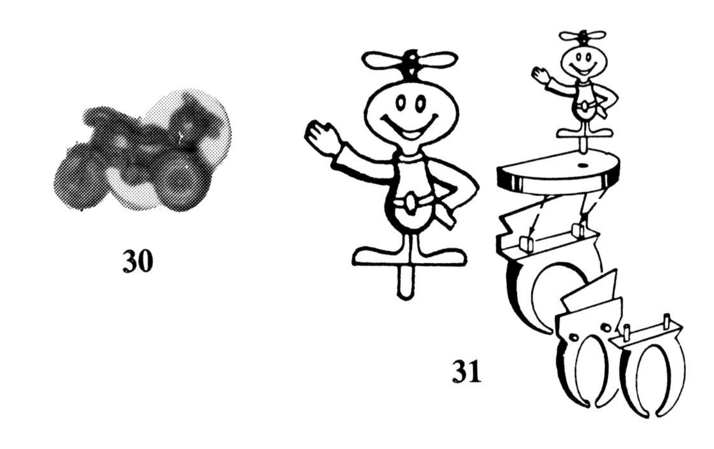

30 31

30) QUISP FIGURE ON UNICYCLE (Quaker 1969) Snap-together blue plastic 2.5"tall detailed figure of Quisp riding a working unicycle which is powered by a small ball under the frame. This premium came free inside boxes of Quisp cereal. **$50-75**

31) QUISP FRIENDSHIP RING (Quaker 1966) Plastic snap-together ring features a 1" detailed figure of Quisp waving. **$25-50**

32) QUISP "GYRO-POWERED" FIGURE ON THREE-WHEELER (Quaker 1969) Snap-together blue plastic 2.5" long three-wheeler with detailed figure of Quisp riding. The vehicle has a white gear-edged disc and plastic pull-string which, when pulled, sends it moving in a forward direction. This premium was free inside boxes of Quisp cereal. **$35-65**

33) QUISP RAY GUN RING (Quaker mid-60's) 1" long plastic ray gun sits on ring band with elevated base. There is side stem lever behind the back of gun which, when pulled back and released, taps gun, causing a small bullet to eject from barrel. **$50-150**

34

34) TRIX RABBIT BOWL AND MUG SET (General Mills 1963) Plastic white bowl and mug with color accents. The bowl has embossed colored pictures of the Trix rabbit in a variety of predicaments. The mug has two ear-shaped handles and a face on one side. This set was a mail order premium available through Trix cereal coupons. **$20-30**

35

35) TWINKLES CEREAL BOXES (General Mills 1960-61) Twinkle cereal boxes were unique in that each box had a 3-page full color art story on the back of the box featuring Twinkles the Elephant. Cutting along the outline of the first page on the back of the box would reveal the fold-open second and third page. EACH: **$75-150**

36

36) TWINKLES CEREAL BOXES WITH BULLWINKLE (General Mills 1962-63) Before Bullwinkle became Cheerio's official mascot in 1965, he and other Jay Ward characters from the Rocky and Bullwinkle show appeared on the back of Twinkles cereal boxes in the form of puzzles, games, and various activities. Like the earlier Twinkle boxes, the trademark 3-page story back remained. EACH: **$100-200**

37

38

37) TWINKLES AND SANFORDS BOAT BOOK (Top Top Tales 1961) 6"x8" hardback book with 28-page story accompanied by full color story art on each page. **$12-18**

38) TWINKLES AND HIS PALS CHILDREN'S RECORD (Golden Records 1961) 7"x8"stiff cardboard sleeve contains 78 rpm records with songs "The Hide and Seek Game" and "The Parade." **$15-25**

39

41

39) TWINKLES "HIS TRIP TO THE STAR FACTORY" GAME (Milton Bradley 1961) Twinkles the flying elephant was one of the very first cereal box characters to have his own cartoon show and merchandising line. He set the pace for other cereal companies to create their own unique mascot characters on countless brands of cereal. The Twinkles game was the first commercial item produced on Twinkles and sold moderately well. The object of the game was to be the first player to collect the most stars before Twinkles arrives at the star factory. Game includes five cardboard die-cut character head playing pieces, magic stars, board and dice. **$75-125**

40

40) TWINKLES & FRIENDS-KING LEONARDO & HIS SHORT SUBJECTS WASTE PAPER BASKET (1960) Colorful tin litho child's waste can stands 12" high and features Twinkles and his friends on one side, including two other General Mills cereal mascots (The Trix Rabbit and the Frosty-0 Bear). The other side features King Leonardo and six of his subjects. King Leonardo is holding a General Mills banner. **$100-150**

41) TWINKLES SPONGE FIGURES (General Mills 1960) Set of four 5"x3"thin color sponges with illustration of a character on each sponge (Twinkles the Elephant, Fulton the Camel, Wilbur the Monkey and Sanford the Parrot). The figures were to be cut out and placed in water, which would "puff" the sponges up to make the characters three dimensional. The trademark of Twinkles cereal was the free comic book on the back of the box so very few premiums were offered inside the cereal. This premium was a mail order item available through Twinkles cereal. **$25-40**

COMIC STRIP CHARACTERS

Before network television completely dominated the cartoon kingdom, characters appearing in the newspapers and comic books held steady ground in merchandising. King Features Syndicate, Inc. was among the largest of the newspaper comic strip licensors (Steve Canyon, Beetle Bailey, Flash Gordon, The Phantom, Snuffy Smith, Popeye, Blondie) and waves of toys on King's characters were produced over the decades prior to the late Sixties.

1

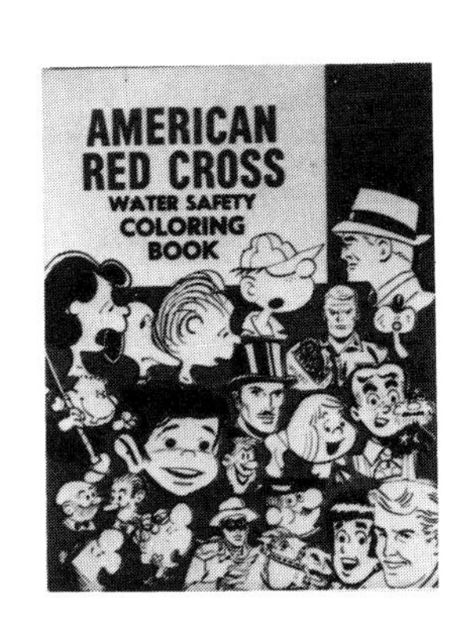

2

4

5

1) ALLEY OOP COLORING BOOK (Treasure Books 1962) 8"x11" 50-pages. **$15-25**

2) AMERICAN RED CROSS SAFETY COLOR BOOK (1970) Dick Tracy, Peanuts, Archie, Mandrake, Beetle Bailey, Moose, BC and more. **$15-30**

4) BEETLE BAILEY BENDABLE FIGURE (Toy House 1963) 2.5" bendable rubber figure of Beetle comes on 6x4" display card with working compass. **$20-25**

5) BEETLE BAILEY BOBBING HEAD FIGURES (Lego 1961) 8" composition figures with spring-mounted head. Beetle Bailey, Sgt. Snorkle, Lt. Fuzz, and Zero. EACH: **$75-125**

3

6

3) BEETLE BAILEY BELT W/BUCKLE Child's cloth belt with Beetle's name sewn in the whole length of the belt. The metal buckle features a color illustration of Beetle and Sgt. Snorkel. **$10-15**

6) BEETLE BAILEY BOARD GAME (Jaymar 1956) Included in this game are miniature plastic Marx soldiers, exclusive to this game only. **$50-65**

7

7) BEETLE BAILEY CANDY BOXES (Phoenix Candy 1964) 3"x4"x1"cardboard box features Beetle Bailey on front and color by number illustration on back. Series of four made (Beetle Bailey, Sarge, Zero and Lt. Fuzz). EACH: **$25-35**

8

9

8) BEETLE BAILEY COLORING BOOK (Lowe 1961) 8"x11",100-pages. **$15-25**

9) SUNDAY FUNNIES COLA "PRESENTS BEETLE BAILEY" CAN (Flavor Valley Corp. 1968) 5" pull-top cola can features colorful graphics of Beetle Bailey standing with mop pail beside a two-panel newspaper strip. Part of a series of eight cola cans by Flavor Valley Corp. on King Features Syndicate comic strip characters. **$25-50**

10

10) BEETLE BAILEY "ZERO" HAND PUPPET (Gund 1960) 11"soft vinyl head/cloth body puppet of Beetle's friend Zero. Comes in 8"x5"x4"window display box. **$40-50**

11

11) BEETLE BAILEY "FOLD-A-WAY" CAMP SWAMPY PLAYSET (MPC 1964) 12"x20"x3"deep box folds open into an army camp grounds and comes with cardboard barracks, gate with Camp Swampy overhang, plastic soldiers, jeeps, trucks, exploding bridge and five 3" detailed character figures of Beetle, Sarge, Killer, Zero and General Halftrack. **$150-200**

12

13

12) BEETLE BAILEY JIGSAW PUZZLE (Jaymar 1963) Colorful 7"x10"box contains 60-piece puzzle which assembles into a 14"x10"scene of Beetle Bailey and Sarge. **$15-25**

13) BEETLE BAILEY MAGIC DRAWING SLATE (Lowe 1963) 8"x12"illustrated cardboard display card holds lift-up erasable film sheet and comes with wood stylus. **$25-35**

14

14) BLONDIE AND DAGWOOD INTERCHANGEABLE FACE BLOCKS (Gaston Mfg. 1951) 7"x9"box contains wood block pieces, each with all four sides painted a different facial feature. Over one trillion variations can be arrived at. **$25-50**

15) BLONDIE BOARD GAME (Parker Brothers 1969) Dagwood and family on box. **$15-20**

16) BLONDIE BOARD GAME (TransOgram 1966) **$20-25**

17) BLONDIE CRAYONS (1952) 5.5"x4"x2"box of crayons. **$20-30**

18) BLONDIE JIGSAW PUZZLE (Jaymar 1963) 17"x10"box contains 100-piece puzzle which assembles into a 22"x17" scene of Dagwood and family. **$15-25**

19

19) BLONDIE LUNCHBOX WITH THERMOS (1968) Box: **$75-100** Thermos: **$20-30**

20

20) BLONDIE VIEWMASTER PACKS (1960) 5"x5"color illustrated envelope contains three reels and 16-page story booklet. EACH: **$10-20**

21

21) DAGWOOD SANDWICH BAG (1952) 7"x5"illustrated bag of Dagwood loaded down with armfulls of sandwich making goodies. **$15-25**

22

22) BUCK ROGERS BOARD GAME (Transogram 1965) Colorful box graphics depict Buck Rogers fighting Alien creatures. Object of the game is for players to land on each of six planets to pick up a delegate while avoiding dangerous orbiting comets. Game includes colorful 8"diameter Universe Disc, six die-cut stand-up space-kid figures, spaceships, comets and color illustrated cards. **$50-100**

23

24

23) **CAPTAIN AND THE KIDS COLORING BOOK** (1950's) 8"x11",80+ pages. **$20-25**

24) **DENNIS THE MENACE COLORING BOOK** (Watkins-Strathmore 1960) 8"x11",80+ pages. **$10-20**

25

25) **DENNIS THE MENACE CARTOON KIT** (Colorforms 1960) 12"x14" large deluxe-size edition set of thin vinyl plastic character pieces that affix to illustrated backgroud board. **$30-40**

26) **DENNIS THE MENACE DELUXE SIZE COLORFORMS SET** (1960) Large 15"x20" box contains die-cut thin vinyl stick-on character pieces and accessory pieces, plus background scene for placement of characters, and fold-out instruction booklet. **$25-35**

27

27) **DENNIS THE MENACE MISCHIEF KIT** (Hasbro 1955) 8"x5" fold-open box an assortment of "harmless, hilarious pranks" which include a rubber lizard, rubber bug, fake sugar cubes, party horn and finger trap device. **$25-50**

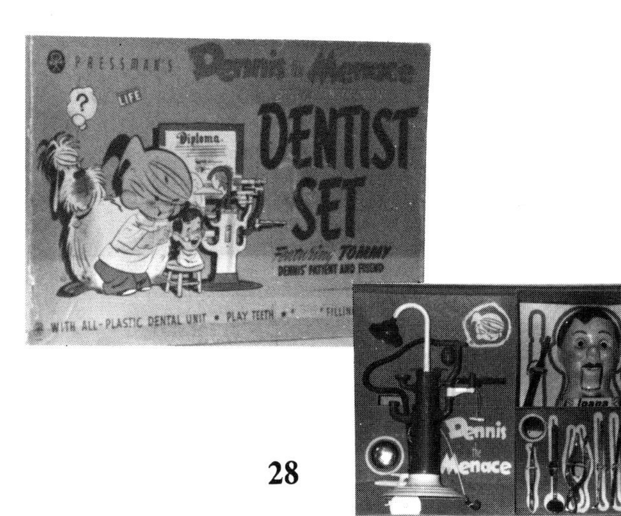

28

28) **DENNIS THE MENACE DENTIST SET** (Pressman 1950's) Large 22"x16"x6" deep box contains intricate plastic dental unit, seven dental instruments, water bowls, filling material, toothbrush, tube of Ipana toothpaste, and a composition bust of a patient. The mouth of the bust can open to revel several teeth which require dental attention. **$100-150**

29

30

29) **DENNIS THE MENACE HEAD-SHAPED MUG** (Kelloggs 1960) 3.5"tall hard plastic mug molded in the shape of Dennis's head and wearing a red western hat and scarf. Face has color accents. Originally offered with cereal bowl (not shown) by Kellogg's Corn Flakes. MUG: **$10-15** BOWL: **$8-12**

30) **DENNIS THE MENACE THEME SONG RECORD** (Golden 1960) 7"x8" illustrated stiff cardboard sleeve holds 78 rpm record which contains TV theme song and "Ka-pow, Ka-pow, Ka-pow" song. **$8-12**

31

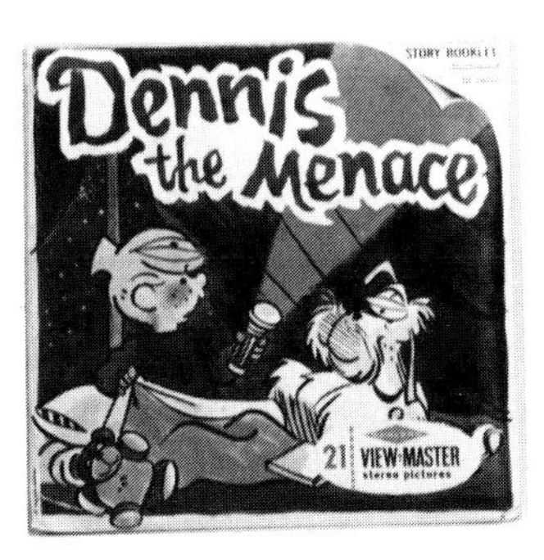

32

31) **DENNIS THE MENACE TIDDLY WINKS GAME** (Whitman 1960) **$15-20**

32) **DENNIS THE MENACE VIEWMASTER REEL SETS** (Sawyer 1961) 5"x5"color illustrated envelope contains three reels and 16-page story booklet. EACH: **$12-18**

33

35

33) **DENNIS THE MENACE AND RUFF VINYL FIGURE LAMP** (Hall Syndicate 1960) 7" painted molded vinyl figures of Dennis the Menace and Ruff with light switch and cord in back. **$50-80**

34) **DONDI COLORING BOOK** (Saalfield 1960) Based on 1960 movie with David Janssen. 8"x11",80+ pages. **$15-25**

35) **DONDI JIGSAW PUZZLE** (Jaymar 1961) 7"x10"box contains 60-piece puzzle which assembled into a 10"x15"scene of Dondi swimming at summer camp. The scene is titled "Water Fun." **$15-25**

36

36) **DONDI PENCIL CASE** (Hasbro 1961) 4"x8"x1"red cardboard box with slide open drawer. Illustrated color paper decal features Dondi pulling a wagon as his dog rides inside. **$15-30**

37

37) **DONDI FINDERS KEEPERS GAME** (Hasbro 1960) 10"x18"box contains playing board, playing pieces and spinner. **$25-40**

38

38) **DONDI POTATO RACE GAME** (Hasbro 1960) 10"x18" box contains playing board, pieces and spinner. **$35-50**

39

39) DONDI "PRAIRIE RACE" BOARD GAME (Hasbro 1960) 18"x10"box contains playing board, playing pieces and spinner. **$25-40**

40

41

40) DONDI WONDER BOOK (1960) 7"x8"hardback book with 28-page story accompanied by full color story art on each page. **$5-10**

41) FLASH GORDON AND THE MARTIAN MODEL KIT (Revell 1965) 6"x10" box contains 1/10 scale all plastic assembly kit of Flash Gordon, a martian and moon crater style base. **$100-200**

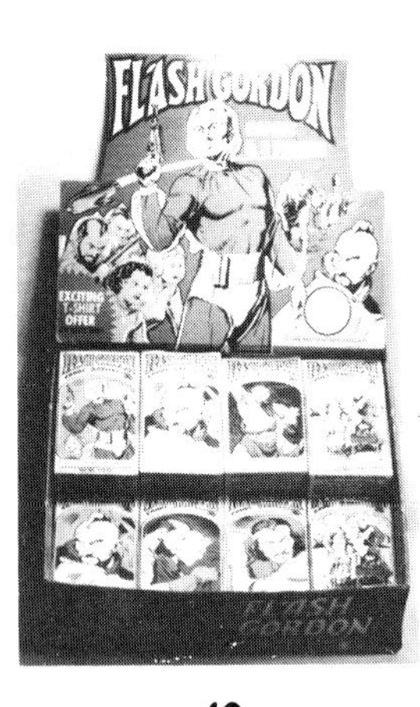

42

43

42) FLASH GORDON CANDY BOX STORE DISPLAY (1970s Phoenix Candy) Large colorful display box with pop open lid that displays 16 boxes of Flash Gordon candy with trading card on back. **$50-100**

43) FLASH GORDON COLORING BOOK (Whitman 1952) 8"x11",100-pages. **$25-50**

44

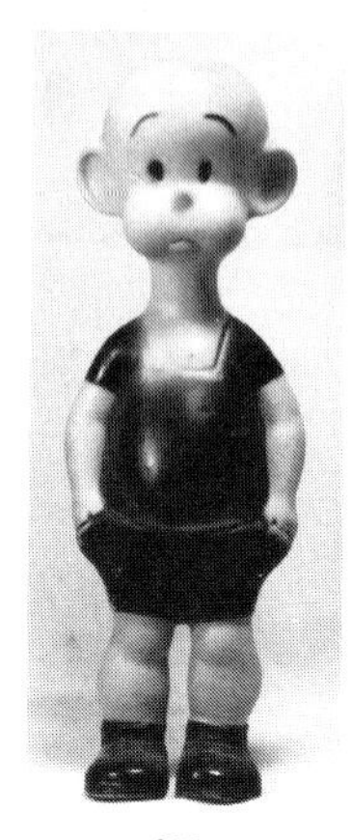

45

44) FLASH GORDON-HIS ADVENTURES IN SPACE -- A BOOK TO COLOR (Artcraft 1965) 8"x11",50-pages. **$20-25**

45) HENRY VINYL FIGURE (1950's) 9" tall hollow soft vinyl figure with painted accents. **$25-50**

46

47

46) HENRY IN LOLLIPOP LAND BOOK (Wonder Books 1950's) 7"x8"hardback book with 20+ page story with full color story art on each page. **$8-12**

47) KRAZY KAT MAGIC DRAWING SLATE (Lowe 1963) 8"x12"illustrated cardboard display card holds lift-up erasable film sheet and comes with wood stylus. **$25-50**

48

48) LI'L ABNER BOARD GAME (MB 1942) **$75-100**

49

49) LI'L ABNER PAINT SET (Gem Color Co. 1948) 16"x11" box contains 26 circular paint tablets affixed to illustrated display inlay, two water trays, six crayons, brush and 5"x5"color booklet. **$50-75**

50

51

50) LI'L ABNER GAME (Parker Bros. 1969) 18"x10"box contains colorful cartoon cards of the comic strip, playing pieces, playing board, direction cards, and dice. Object is to be the first player to reach Sadie Hawkins. **$25-50**

51) LI'L ABNER "SHMOOS" SALT & PEPPER SHAKERS (1950's) Ceramic pair of Shmoos. **$25-50**

52

53

52) MISS PEACH CHILDREN'S RECORD (RCA 1960) 45 rpm record comes in illustrated paper sleeve. **$10-12**

53) MISS PEACH PAPERBACK (Pyramid 1962) **$8-12**

54

55

54) NANCY & SLUGGO GIANT COLORING BOOK (Saalfield 1950) 15"x9"book. **$25-75**

55) NANCY & SLUGGO BOARD GAME (MB 1944) **$150-250**

56

56) NANCY & SLUGGO DOLLS (S&P Doll and Toy Co. 1954) 16"x6"x5"deep box contains 16" tall doll with molded vinyl plastic head with realistic eyes and painted accents and stuffed soft vinyl body with cloth outfit. Nancy is dressed in a red and green plaid skirt, white blouse, black vest, white socks and soft vinyl plastic shoes. Sluggo is dressed in blue pants, red and white striped shirt, black jacket and soft vinyl red shoes. Box lid is designed as a three paneled comic strip featuring Nancy and Sluggo. (Note: Nancy is actually labeled on the box as "Sluggo's Girlfriend.") NANCY: **$150-200** SLUGGO: **$150-200**

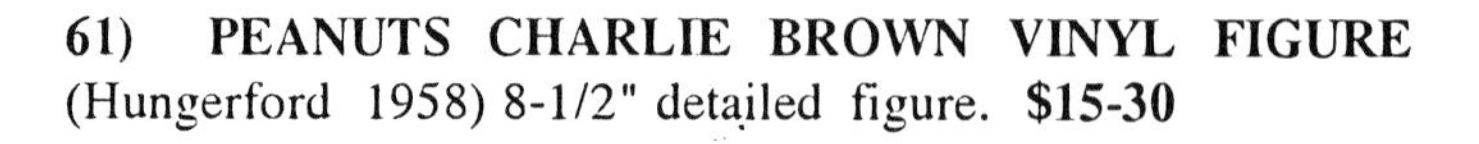

57) SLUGGO DOLL (Knickerbocker 1971) Small 8" doll comes n 8"x4"display box. **$15-25**

58

59

58) PARADE OF COMICS PASTE-UP BOOK (1960) Large color cut-out book includes Phantom, Dondi, Prince Valiant, Alley Oop, Beetle Baily and more. **$25-30**

59) PEANUTS--A BOOK TO COLOR (Saalfield 1965) 8"x11",60-pages. **$8-15**

60

60) PEANUTS BOARD GAME (S&R 1959) 10"x20"box contains playing board, four playing pieces, 20 illustrated tiles and dice. The object of the game is to be the first player to form complete sequences of card tiles. **$25-50**

61

61) PEANUTS CHARLIE BROWN VINYL FIGURE (Hungerford 1958) 8-1/2" detailed figure. **$15-30**

62

64

62) PEANUTS COLORING BOOK (Saalfield 1953) Colorful cover designed with Snoopy heads on the spine. First Peanuts coloring book and one of the first Peanuts items made. **$35-75**

63) PEANUTS COLORING BOOKS (Saalfield/Artcraft 1960's) EACH: **$8-12**

65

66

64) PEANUTS FIGURAL SHAMPOO BOTTLES (Avon 1969) Pair of 6" painted plastic figural bottles of Charlie Brown and Lucy. Bottle caps are disguised as a baseball cap and beanie. EACH: **$8-12**

65) PEANUTS LUCY'S TEA PARTY GAME (MB 1971) Large 13"x18"x4"deep box contains illustrated vinyl playing sheet, plastic serving tray, four plastic tea cups, tea pot and plastic sugar cubes. Game is played with water and the object of the game is to be the first player to have his/her tea cup overflow. **$20-30**

66) **PEANUTS MAGIC SLATE** (Saalfield 1961) 8"x12" illustrated cardboard display card holds lift-up erasable film sheet and comes with wood stylus. **$20-30**

67

67) **PEANUTS POCKET DOLLS** (Boucher 1968) Set of five 7" plastic dolls with cloth outfits. Linus is complete with detachable blanket, Charlie Brown is wearing baseball cap, Snoopy has his flying helmet with goggles, and Schroeder's sweatshirt has an illustration of Beethoven on it. SET: **$75-100**

68) **PEANUTS SNOOPY ASTRONAUT PLAYSET** (Knickerbocker 1969) Large boxed set includes a 5" vinyl Snoopy, spacesuit, helmet, moon shoes, camera, life-support system, U.S. flag, large moon rover and 4" vinyl figure of Woodstock. **$25-50**

69

69) **PEANUTS "SNOOPY" GAME** (Selchow & Righter 1960) 10"x20"box contains 16 cardboard discs, each with an illustration of a different breed of dog, Snoopy disc, spinner and playing board. Object is to get each dog back to his home. **$40-50**

70

70) **PEANUTS "STUFF-N-LACE" DOLLS** (Standard Toykraft 1961) 14"x9"window display box contains two cloth Peanuts characters, stuffing and lace. Two sets were produced, one featuring Charlie Brown and Lucy, the other featuring Linus and Snoopy. **$35-50**

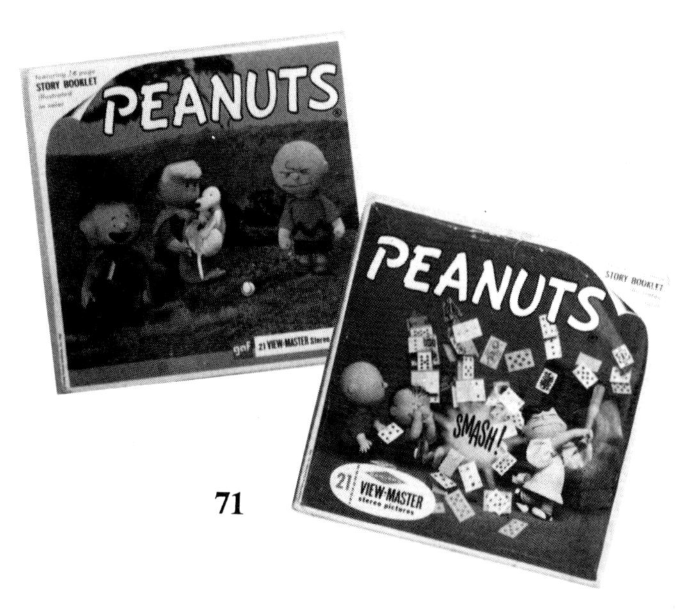

71

71) **PEANUTS VIEWMASTER REEL SETS** (Sawyer 1960) 5"x5"color illustrated envelope contains three reels and 16-page story booklet. EACH: **$12-18**

72 **73**

72) **PEANUTS & SNOOPY MAGIC DRAWING SLATE** (Saalfield 1967) Colorful die-cut cardboard with attached draw-and-lift film. Illustrations of several characters on front and puzzles and games on back. **$20-25**

73) **CHARLIE BROWN BOBBING HEAD FIGURE** (Lego 1961) 7" composition figure with spring mounted head. **$35-60**

74

75

78

74) PIG PEN BOBBING HEAD FIGURE (Lego 1961) 7" composition figure with spring mounted head. **$35-60**

75) GOOD GRIEF CHARLIE BROWN: PEANUTS RECORD ALBUM (Columbia 1963) **$10-20**

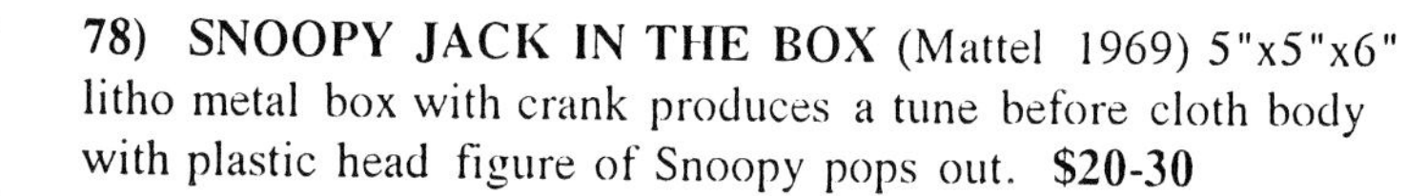

78) SNOOPY JACK IN THE BOX (Mattel 1969) 5"x5"x6" litho metal box with crank produces a tune before cloth body with plastic head figure of Snoopy pops out. **$20-30**

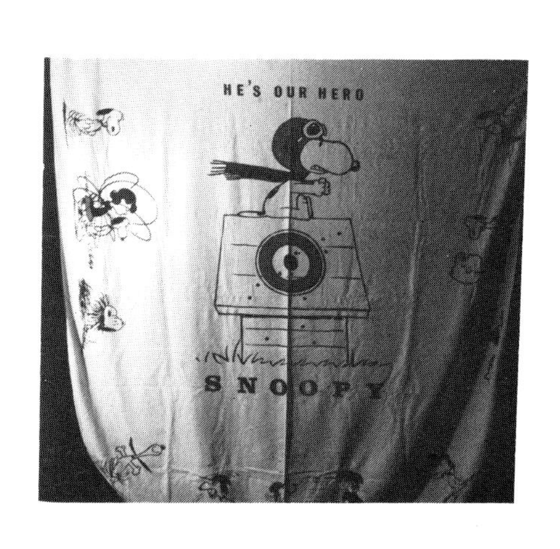

76

77

79

76) HIDE-A-SNOOPY (Aviva 1964) 12"blue cone has wood rod which pushes a hidden 8"stuffed cloth Snoopy to surface. Cute and uncommon with tag. **$15-30**

77) SNOOPY BEDSPREAD (1968) Large quality fabric twin-size bedspread with black fabric trim. Center illustration features Snoopy as the WWI Flying Ace with the caption above reading, "He's Our Hero." There are several smaller illustrations of Snoopy in various classic scenes such as sleeping on his doghouse, skipping rope with Lucy, playing baseball with Linus, etc. **$50-75**

79) SNOOPY AND THE RED BARON GAME (MB 1970) Deluxe size game consisting of cardboard and plastic pieces that construct into an action diorama of Snoopy on his doghouse in battle with the Red Baron, with background walls illustrated with the Peanuts gang looking on. Object of the game is to shoot marbles at Snoopy and Snoopy only tries to catch the good ones. **$25-35**

80

81

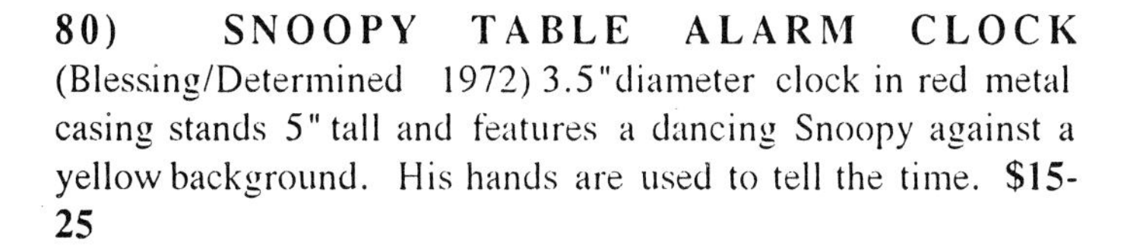

80) SNOOPY TABLE ALARM CLOCK (Blessing/Determined 1972) 3.5"diameter clock in red metal casing stands 5" tall and features a dancing Snoopy against a yellow background. His hands are used to tell the time. **$15-25**

81) SNOOPY TABLE ALARM CLOCK (Equity 1968) 4" diameter clock in black metal casing stands 5" tall and features Snoopy playing tennis against a yellow checkerboard background. His hands are used to tell the time. **$20-30**

82

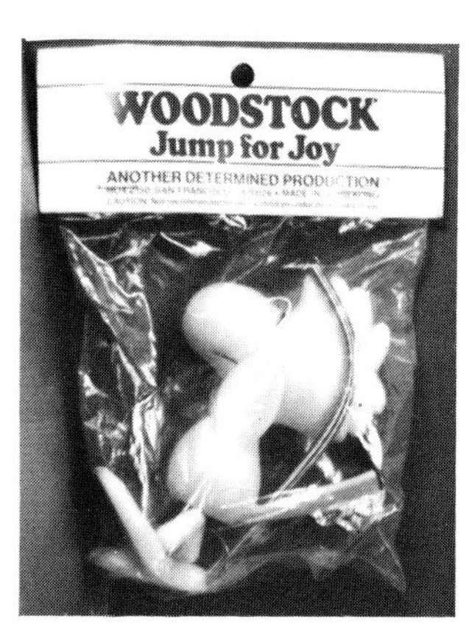

83

82) SNOOPY "WORLD'S GREATEST TENNIS PLAYER" TROPHY (Aviva 1970) 5" figure of Snoopy on trophy base. **$8-15**

83) WOODSTOCK "JUMP FOR JOY" FIGURE (Determined Prod. 1971) 5" yellow plastic figure of Woodstock features a coiled spring on the top of his head and movable leads. As the spring is dangled, Woodstock appears to be hopping and walking. **$12-18**

84

84) POGO CHARACTER FIGURES (Duz Detergent Soap 1969) Sculpted 4"-5"vinyl plastic figures of Walt Kelly's Pogo comic strip characters. Colorfully painted with movable heads and/or arms. Originally given as a premium with boxes of Duz Detergent Soap. EACH: **$15**

- **A) Porky Pine**
- **B) Albert Alligator**
- **C) Pogo**

85

85) POGO MOBILE (1950's) 10"x14"illustrated stiff paper envelope contains 22 cut-out die-cut cardboard figures of Pogo and supporting character. **$75-125**

86

86) PRINCE VALIANT GAME OF VALOR (TransOgram 1955) x18" box contains playing board, spinner, a chart of Prince Valiant's deeds, two merit point boards and 12 color plastic markers. Object of the game is to pass through the gateway, travel around the red moat path, escaping dangerous perils, and being the first player to enter the castle square. **$25-75**

87

88

87) PRINCE VALIANT CHILDREN'S BOOK (Little Golden 1955) **$8-15**

88) PRINCE VALIANT CROSSBOW PISTOL GAME (Parva Products Co. 1948) 9"x10"box contains plastic and wood spring-loaded crossbow pistol, three rubber tipped arrows, and two colorful cardboard targets, one of Prince Valiant's shield and one of a wild boar. **$50-75**

89

89) PRINCE VALIANT FIGURE (Schucco 1942) Detailed 6" pressed wood figure on base. **$100-200**

90

90) SNUFFY SMITH "TIME'S A-WASTIN'" BOARD GAME (MB 1963) 8"x16"box contains playing board, four playing pieces and spinner. Object of the game is to be the first player to move all of his pieces from start to finish. **$25-50**

91) SNUFFY SMITH JIGSAW PUZZLE (Jaymar 1963) 9"x13"box which contains 100-piece jigsaw puzzle which assembles into a 22"x17"scene of Snuffy Smith and gang. **$15-25**

92

92) STEVE CANYON BOARD GAME (Lowell 1959) Comes with four separate stand-up cockpit/control panels with operationg gauges, etc. Cards include weather bulletins to help the pilots fly more safely. Box is 10"x20". **$50-75**

93

93) STEVE CANYON COLORING BOOK (Saalfield 1952) Oversized 11"x15"contains b/w comic strip story illustrated by Milton Coniff. **$50-75**

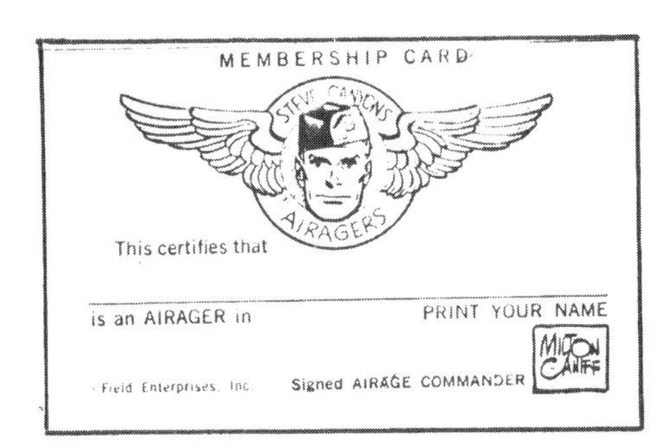

94

94) STEVE CANYON FLIGHT MEMBERSHIP IDENTIFICATION CARD (1959) Illustrated 1"x3"card has place for your photo. **$5-10**

96

97

95) STEVE CANYON HALLOWEEN COSTUME (Halco 1959) 10"x12"box contains mask and one-piece fabric bodysuit with illustration of Steve Canyon on front. **$25-50**

96) STEVE CANYON JET HELMET (Ideal 1959) Colorfully illustrated 12"x12"box contains well-made white plastic helmet with green shades, visors and decals on front of helmet. **$50-100**

97) STEVE CANYON LUNCHBOX WITH THERMOS (1959) Box: **$100-200** Thermos: **$25-35**

98

100

99

98) STEVE CANYON PILOT'S CAP (1959) Blue fabric officer's flight hat with stiff black visor and strap. Front of cap reads "Steve Canyon" in white letters. **$20-40**

99) STEVE CANYON MODEL KIT (Aurora 1959) 5"x13" box contains 1/8 scale all-plastic assembly kit depicting Steve Canyon in jet fighter uniform. **$100-200**

100) STEVE CANYON SCHOOL BAG (1959) Large colorful illustrated yellow canvas bag. **$20-30**

101) STEVE CANYON STENO PAD NOTEBOOK (Hytone 1958) Colorful cover depicts several scenes of Steve and his friends. Milton Caniff artwork. **$5-15**

102) STEVE CANYON T-SHIRTS (1959) Came with a membership card and gold Flight Wings (nice tin litho). **$25-50**

101

102

DICK TRACY

"The Dick Tracy Show" premiered as a half-hour cartoon series in 1960 and featured Dick Tracy with a newly created crew of supporting law enforcers including Hemlock Holmes, Jo Jitsu, Go Go Gomez and Heap O'Callory. Despite a large merchandising campaign (1960-1962), the series fell out of syndication. Merchandising was revived in 1967 when a television series starring Ray MacDonnald was filmed, but never aired.

1

1) DICK TRACY CANDY BAR WRAPPER (Schutter Candy Co. 1950's) 5"x3" candy bar sized wrapper features illustration of Dick Tracy against a bright blue background with red lettering. **$15-25**

3

3) DICK TRACY "CRIME STOPPER" GAME (Ideal 1963) Huge plastic computer control board panel is designed with clue cards that hide the face of the criminal the player is trying to identify. Comes in large 28"x18"x6" color box. **$25-75**

2

2) DICK TRACY COLORFORMS SET (1962) 8"x12"box contains die-cut thin vinyl stick-on character -pieces and accessory pieces, plus background scene for placement of characters, and fold-out instruction booklet. **$25-40**

4

4) DICK TRACY "CRIME STOPPER" PREMIUM KIT (1961) Boxed set contains illustrated Dick Tracy wallet, badge with embossed head of Tracy, flashlight, whistle, secret code maker, several "summons" by the crimestopper club and several illustrated Crime Stoppers textbook sheets by Gould. Also a 5x7"b/w photo of Dick Tracy and Jo-Jitso is included. This was a mail order premium. **$25-75**

5

6

8

8) DICK TRACY JUNIOR DETECTIVE KIT (Golden Press 1962) 8"x13"stiff paper punch-out book with six pages of police equipment to be punched out to form complete detective kit. **$20-30**

9

5) DICK TRACY HANDPUPPET (Ideal 1961) 11" handpuppet with molded vinylhead and fabric body. **$50-100**

6) DICK TRACY'S "HEMLOCK HOLMES" HAND PUPPET (Ideal 1961) 11"puppet with detailed soft vinylhead and cloth body. From the cartoon series Dick Tracy. **$50-75**

9) DICK TRACY LUNCHBOX WITH THERMOS (Aladdin 1967) Box: **$100-125** Thermos: **$50-75**

7

7) DICK TRACY JIGSAW PICTURE PUZZLE (Jaymar 1961) 10"x7"box contains 60-piece jigsaw puzzle which, when assembled, measures 10"x14"and depicts Tracy in charge of a police line-up with Flat Top, The Mole and Stooge Villa. **$15-25**

10

10) DICK TRACY MASTER DETECTIVE GAME (Selchow & Righter 1961) 10"x20" box contains four cardboard stand-up characters from the Dick Tracy cartoon, one villain plus 28 playing tiles of stolen loot. Object is to capture villain before he flees country and regain as many stolen loot tiles as possible. Contraband loot cards include drugs, jewels, furs, etc. **$25-50**

11

11) DICK TRACY OIL PAINTING SET (Hasbro 1967) 26"x18" box contains 22"x14" pre-numbered, pre-sketched canvas, ten vials of paint and brush. **$75-125**

12) DICK TRACY PATROL SQUAD GUN SET (Mattel 1961) Set includes 28" long black/brown plastic tommy gun with side bolt action and Official Dick Tracy logo on stock, a metal .38 snub-nose pistol which shoots shells, and shoulder holster. Comes in 32"x14"x3"deep box. Interestingly enough, some stores in New York City and the state of Massachusetts would not carry this gun set! **$200-250**

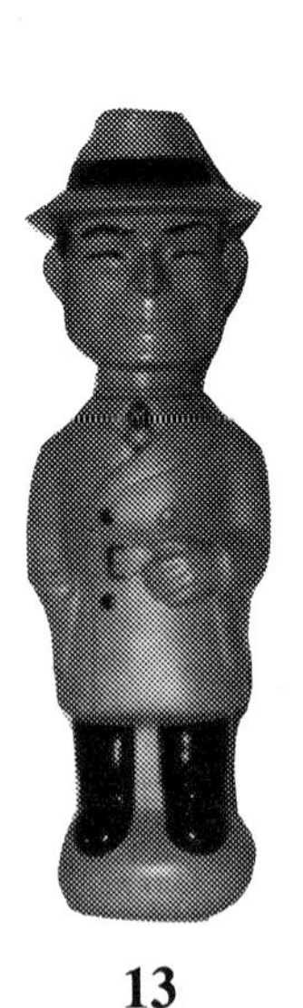

13

14

13) DICK TRACY "SOAKY" BUBBLE BATH CONTAINER (Colgate-Palmolive 1963) 10" plastic soap container with hard plastic removable head. **$25-50**

14) DICK TRACY SPARKLE PAINTS (Kenner 1962) 8"x12"box contains six pre-numbered sketches and five vials of sparkle paint and features characters from the Dick Tracy cartoon such as Jo Jitsu and Go Go Gomez. **$65-100**

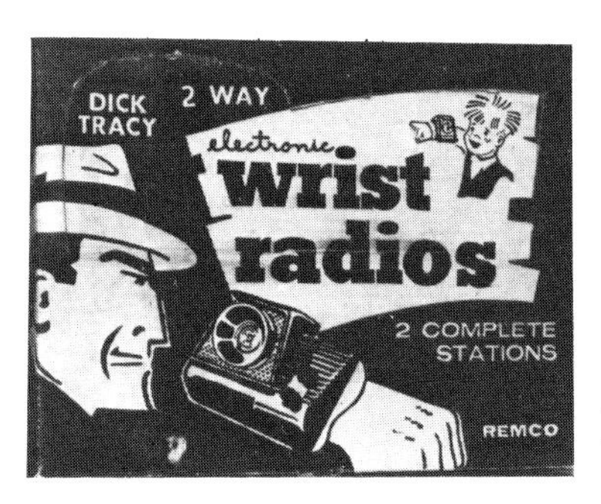

15

15) DICK TRACY WRIST RADIOS (Remco 1961) 9"x13" box contains two battery-powered plastic wrist-radios about 2"x3.5"x1"in size. A variety of wrist-radios were made during the Sixties, many by Remco. **$50-100**

16

16) DICK TRACY SPACE COUPE MODEL KIT (Aurora 1967) 13x5"box contains 1/72 scale all-plastic assembly kit of police space coupe and comes with four H/O scale character figures. **$50-100**

17

17) DICK TRACY MODEL KIT (Aurora 1967) 13x5"box contains 1/16 scale all-plastic assembly kit which depicts Tracy climbing a fire escape with pistol in hand. **$75-125**

DR. SEUSS CHARACTERS

Like the newspaper comic strip "Peanuts" by Charles Schultz, the Dr. Seuss children's book characters began rising in popularity in the late 50's and flourished throughout the Sixties and Seventies. Several half-hour cartoon specials were produced on the Seuss characters over the years, and the classic "How the Grinch Stole Christmas" remains a network holiday tradition.

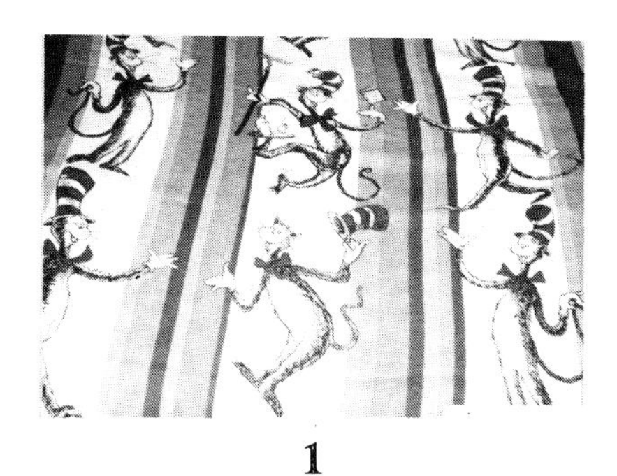

1

1) **DR. SEUSS CAT IN THE HAT BED SPREAD AND CURTAIN SET** (1969) Two pair of curtains and matching bed spread. **$25-50**

2

3

2) **DR. SEUSS CAT IN THE HAT MODEL KIT** (Revell 1959) 6"x10" box contains all plastic assembly kit of Dr. Seuss's Cat in the Hat which, when assembled, stands over 9" tall. Includes brush and bottle of cement glue. **$50-100**

3) **DR. SEUSS CAT IN THE HAT WITH THING 1 AND THING 2 MODEL KIT** (Revell 1960) 7"x10"box contains all plastic assembly kit of the Cat in the Hat plus two small figures of Thing 1 and Thing 2. This kit features easy snap-together parts, turnable heads on all characters, plus the Cat in the Hat's hand, arms and bowtie move. **$150-250**

4) **DR. SEUSS CAT IN THE HAT TALKING DOLL** (Mattel 1970) Well-made 20" stuffed doll with vinyl head and hat. **$100-150**

5) **DR. SEUSS CAT IN THE HAT TALKING HAND PUPPET** (Mattel 1969) 20" tall cloth body with molded vinyl head has voice-activated pull string which, when pulled, produces eleven different phrases. **$50-75**

6

6) **DR. SEUSS CAT IN THE HAT VINYL CARRY BAG** (1965) 13"x16" yellow soft vinyl plastic bag with colorful illustration of the Cat in the Hat on front. **$20-25**

7

7) **DR. SEUSS HEDWIG BIRD TALKING DOLL** (Mattel 1970) 14" long stuffed plush doll with molded vinyl facial features and beak has voice activated pull string which, when pulled, produces eleven different phrases. **$75-125**

8

10

12

8) **DR. SEUSS CAT IN THE HAT JACK IN THE BOX** (Mattel 1970) 5"x5"x6"litho metal box with crank handle produces a tune before figure of the Cat in the Hat pops out. **$50-60**

9) **DR. SEUSS NORVAL THE BASHFUL BLINKET MODEL KIT** (Revell 1959) 6"x10"box contains all plastic snap-together assembly kit of the Dr. Seuss character Norval which when assembled stands over 8" tall. **$50-75**

10) **DR. SEUSS LUNCHBOX WITH THERMOS** (1970) Box: **$50-75** Thermos: **$15-25**

11

11) **DR. SEUSS RECORD ALBUMS** (RCA 1966) Four different Dr. Seuss record albums were produced in the 1960's with RCA Records. Each is 33-1/3 rpm record with two or more Dr. Seuss stories. EACH: **$10-20**

12) **DR. SEUSS THE BIRTHDAY BIRD MODEL KIT** (Revell 1960) 8"x10"box contains all plastic assembly kit of the Birthday Bird carrying birthday cake. Also comes with an assortment of plastic letters to place upon the birthday cake. **$100-200**

13

13) **DR. SEUSS "YERTLE THE TURTLE" GAME** (Revell 1959) Great game of balance includes over twenty little Dr Seuss-style turtles! **$75-125**

14) **DR. SEUSS YERTLE THE TURTLE TALKING DOLL** (Mattel 1972) 10" stuffed plush doll with molded vinyl facial features has voice activated pull string which, when pulled, produces eleven different phrases. Cloth/vinyl doll. **$100-150**

FELIX THE CAT

Felix was revamped in 1960 by animation producer Joe Oriolo and was given a new gimmick -- his magic bag of tricks-- which proved very popular. The half hour series was comprised of two episodes which made up one story and usually involved the scheming bald Professor and his bulldog assistant, Rock Bottom, trying to steal Felix's magic bag of tricks. Major merchandising for this cartoon was between 1960 and 1965.

1

1) FELIX THE CAT CHILDREN'S RECORD (Cricket Records 1958) 45 rpm record contains theme song. **$15-20**

3

3) FELIX THE CAT'S "DOWN ON THE FARM" GAME (Built-Rite 1952) **$40-50**

2

2) FELIX THE CAT COLORING BOOKS (Saalfield 1950's) 8"x11",100-pages. EACH: **$25-30**

4

4) FELIX THE CAT FLASHLITE (Bantam-Lite 1960) 5"x7" display card contains 4" plastic rectangular flashlight with color litho paper illustration of Felix riding on rocket. **$25-35**

5

5) **FELIX THE CAT GAME (Milton Bradley 1960)** 8"x16" box contains playing board, spinner and 24 Felix the Cat cards. Object of the game is to spin the illustrated dial and be the first to complete a picture of Felix the Cat. Cards are obtained by matching the pictures on the spinner with those on the playing board. **$35-40**

1968 Re-issue This game was re-issued in 1968 with a different box lid design. **$20-25**

6) **FELIX THE CAT HALLOWEEN COSTUME** (Halco 1960) 10"x12"box contains mask and one-piece fabric bodysuit with illustration of Felix the Cat on front. **$40-60**

7) **FELIX THE CAT LAMPSHADE** (1952) Stiff color paper shade features space scene with Felix riding rocket. **$25-50**

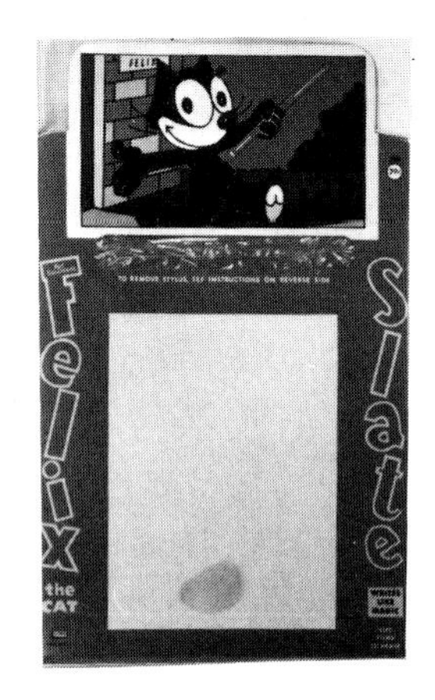

8

8) **FELIX THE CAT MAGIC DRAWING SLATES** (Lowe 1950's) 8"x12" illustrated cardboard display card holds lift-up erasable film sheet and comes with wood stylus. EACH: **$25-50**

9 10

9) **FELIX THE CAT MILK BOTTLE BANK** (1950's) Attractive 15" plastic figure of Felix coming out of a milk bottle. **$50-75**

10) **FELIX THE CAT "MOVIE WHEELS" RECORD ALBUM** (1960) Turning story wheel illustrates what the record is playing. **$10-20**

11

11) **FELIX THE CAT "PASTE WITHOUT PASTE" STICKER BOOK** (Saalfield 1955) Large 11"x12" book features Felix in Candy Land. **$25-50**

12

12) **FELIX THE CAT PENCIL CASE** (Hasbro 1950's) 8"x4"x2"cardboard pencil case with snap-open lid and color litho paper design on lid features Felix, Inky and Dinky. **$25-50**

13

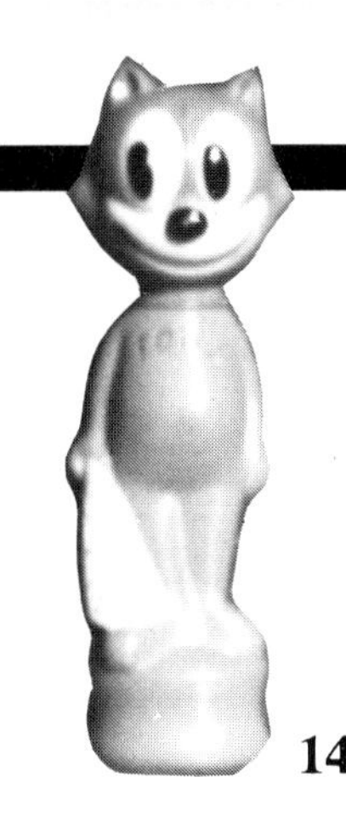

14

13) **FELIX THE CAT PENCIL COLORING BY NUMBERS** (Hasbro 1958) 6"x10"display card holds four pre-numbered sketches of Felix and eight color pencils. **$25-50**

14) **FELIX THE CAT SOAKY BUBBLE BATH CONTAINER** (1963) 11" bubble bath container. **$20-25**

15

15) **FELIX THE CAT STORY BOOKS** (Wonder Books 1960) 7"x8"hardback book with 20+ page story with full color story art on each page. EACH: **$8-15**

16

17

16) **FELIX THE CAT STORE DISPLAY** (Golden 1960) 18" tall heavy die-cut cardboard stand-up store display featuring Felix leaning on a sign post next to his girlfriend Kitty. This display was for children's records. **$100-150**

17) **FELIX THE CAT "SUN-EZE" PRINTING KIT** (Tillman 1962) 3"x4"box contains materials for developing film negatives on paper by light and producing a "Sun-Eze" print. **$15-25**

18

18) **FELIX THE CAT TARGET SET** (Lido 1953) Colorful 10"x14"tin litho target comes in illustrated window display box and includes plastic pistol with rubber tip darts. **$50-100**

19

19) **FELIX THE CAT TILE-SQUARE PUZZLE** (Roalex 1960) 5"x8" display card holds 3"x3" plastic frame which contains 15 illustrated square tiles and one empty tile space. Object is to rearrange tiles to form correct illustration of Felix the Cat. Comes on illustrated 5"x8"card. **$25-30**

20

20) **FELIX THE CAT TRAY PUZZLE** (Built-Rite 1960) Inlay jigsaw puzzle in a 11"x14"frame tray. **$15-25**

GERRY ANDERSON CHARACTERS

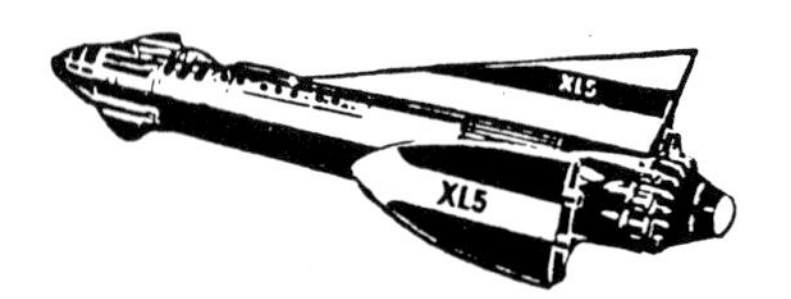

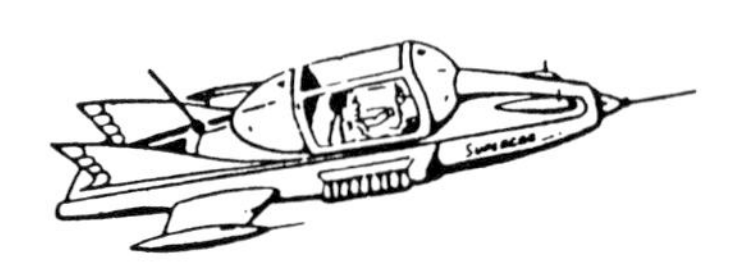

In 1962, British animators Gerry and Sylvia Anderson revolutionized the old-fashioned marionette with "Supermarionation"--a process which employed an electronic "brain" into a plastic marionette which, when in full operation, could blink, shift eyes from side to side and speak in perfect synchronization with a prerecorded soundtrack. Each character was set in a futuristic surrounding and shows were 30 minutes in length.

Supercar debuted in 1962, Fireball XL-5 in 1963, Stingray in 1965, Thunderbirds in 1966, Captain Scarlett in 1967, and Joe 90 in 1968. The three later shows received little U.S. network distribution.

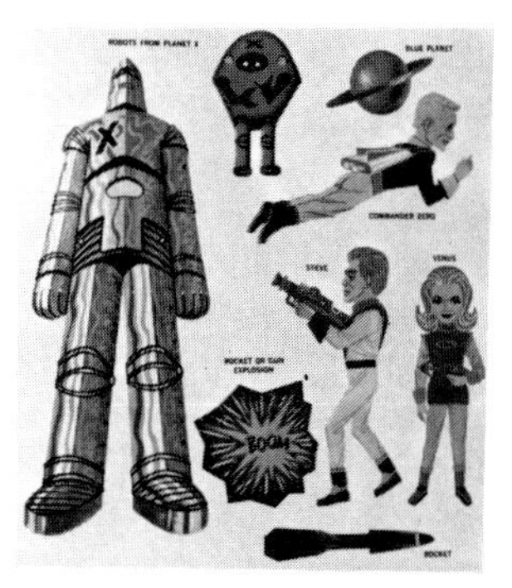
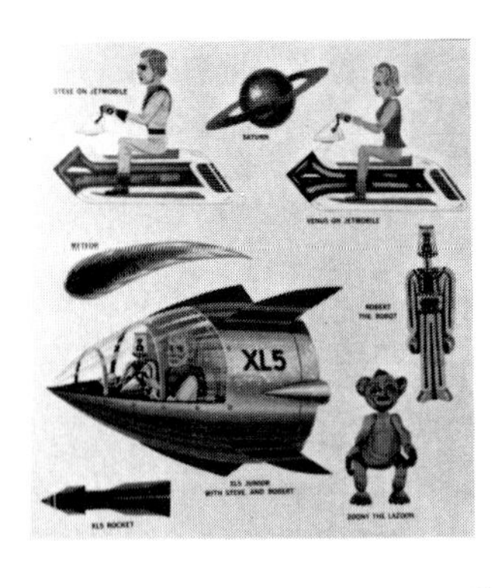

1

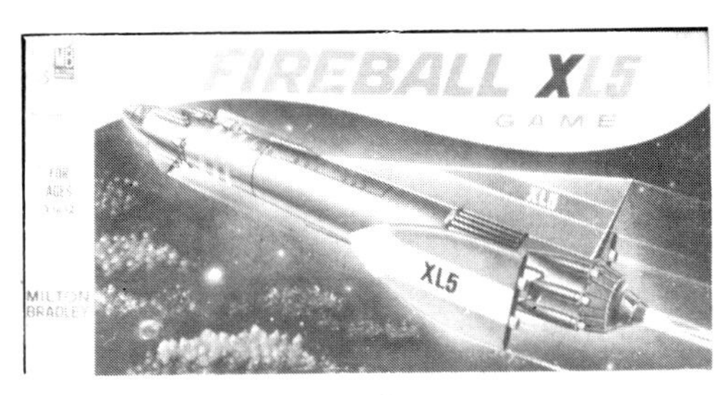

2

2) FIREBALL XL5 BOARD GAME (MB 1963) **$45-60**

3

1) FIREBALL XL5 ACTIVITY SET (Magic Wand 1963) Set contains four 10"x10"color cardboard sheets with punch-out character figures, XL5 spaceship, space sleds, space monster, etc., which can be applied to colorful space background. Box is 10"x14". **$100-150**

3) FIREBALL XL-5 CANDY CIGARETTES (1963) (British distribution) Sweet candy cigarettes. Box is 3"x3". **$25-50**

4) FIREBALL XL5 COLORING BOOK (1963) 8"x11"80+ pages. **$20-25**

5

5) FIREBALL XL5 LUNCHBOX (King Seeley Thermos 1963) Comes with steel thermos which came in two different sizes. Box: **$50-100** Thermos: **$25-35**

6

6) FIREBALL XL5 MAGNETIC DART GAME (Magic Wand 1963) 15"x18"window display box contains colorful tin litho target and four magnetic darts. **$100-150**

7) FIREBALL XL5 MINI PLAYSET (MPC 1963) Colorful window display box contains two levels of figures and space vehicles including character figures. **$100-150**

7

8) FIREBALL XL5 SCHOOLBAG (1963) Colorful 11"x9"x3" canvas, cardboard, and vinyl plastic child's schoolbag features Steve Zodiac, Venus and the Fireball XL5 spaceship on the front. The back has a nice color illustration of the XL5 spaceship leaving its launchramp in Space City with an orbiting moon in background. **$50-100**

9

9) FIREBALL XL5 SPACE CITY PLAYSET (MPC 1964) Large 38"x17"x6"deep box contains 20"long silver plastic XL5 spaceship with detachable nose cone and slide open doors for space sleds and figures, a 36" long working launch ramp, cardboard Space City, eight character figures including two extra figures of Steve Zodiac and Venus which sit on space sleds, plus several accessories including missile and satellite launchers, astronaut figures and space cars. **$500-750**

10

10) FIREBALL XL5 SPACESHIP (MPC 1964) MPC also repackaged the above playset and sold the spaceship with figures in a smaller boxed set which contained all eight character figures and two jet sleds. Box is 22"x10"x10" **$200-300**

11

12

11) FIREBALL XL5 STEVE ZODIAC PUPPET (1964) 5" tall composition and cloth marionette comes in 5"x4"x4"box. British made/distributed. **$200-300**

12) "JOURNEY TO THE MOON" FIREBALL XL5 RECORD ALBUM (Century 21 Records 1963) Soundtrack album contains theme song and story. Color photo album cover. **$50-75**

13

14

13) STINGRAY COLOR PENCIL SET (TransOgram 1966) 8"x10"box contains several pre-numbered sketches and six color pencils. **$50-75**

14) STINGRAY FRAME TRAY PUZZLE (Whitman 1966) Inlay jigsaw puzzle in a 11"x14"frame tray. **$15-30**

15

15) STINGRAY HAND PUPPETS SET (Lakeside 1966) SET OF 4 character puppets; **Troy Tempest, Aquaphibian, X2-Zero, Titan**. Comes in polybag with colorfully illustrated display cards. EACH: **$50-75**

16

16) STINGRAY FRICTION-POWERED SHIP (Lakeside 1966) 12"blue, black, and yellow futuristic looking submarine is friction-powered and features a bobbing periscope when in motion. Rear exhaust is a translucent red plastic. **$350-500**

17

17) SUPERCAR BATTERY OPERATED CAR (Remco 1962) Supercar was the first of British creator Gerry Anderson's puppet animation adventures to be seen by American audiences. Kids fell in love with the entire concept and merchandising soon followed. (A year later Anderson would release Fireball XL5). By far the most elaborate toy made from the show was Remco's rendition of the car. Battery-powered, this 11"orange plastic car was programmed by small directional discs which, depending on the disc used, would drive forward in a figure 8 pattern, loop-to-loop pattern and so on. Four discs were included with the car, each containing two directional programs. There is a full plastic figure of Mike Mercury behind the wheel. The box itself is hollow on the bottom and the car sits on top of the box, protected by a clear plastic dome-like cover. Box measures 10"x12"x6"deep. **$200-300**

18

18) **SUPERCAR CANDY TIN** (Best Plastic, circa 1960's) 5.5"x4"x1"deep heavy tin candy container features Supercar and two space children. British manufactured/distributed. **$25-35**

19

19) **SUPERCAR MAGNETIC TARGET GAME** (Wand 1962) Large 14"x18"colorful tin litho target is built right into the box bottom and comes with 6" long plastic dart space gun on display card and bag of magnetic-tipped projectile darts. **$200-400**

20

20) **SUPERCAR ROAD RACE GAME** (Standard Toykraft 1962) 12"x15"x2"deep window display box contains plastic intricate maze with small tin litho character pieces which are moved about with a magnet stick. Do not be alarmed if you find the tin litho pieces to be Bullwinkle and his gang instead of Supercar character pieces. There was a production error at the Standard Toykraft factory which resulted in almost 30% of these sets containing incorrect pieces! This has little to no effect on the value of this game. **$50-100**

21 22

21) **SUPERCAR SUN-EZE MAGIC PICTURE MAKER SET** (1962) 3"x4"box contains materials for developing film negatives on paper by light and producing a "Sun-Eze" print. **$20-30**

22) **SUPERCAR TO THE RESCUE BOARD GAME** (Milton Bradley 1962) 9"x17"box contains ten cards, spinner, playing board and playing pieces. Object is to be the first player to get to the most disaster areas. **$20-45**

23

24

23) **THUNDERBIRDS COLORING BOOK** (Whitman 1968) 8"x11",80+ pages. **$20-30**

24) **THUNDERBIRDS JIGSAW PUZZLE** (Whitman 1968) Nice 9"x12"color photo boxed puzzle. **$25-30**

25

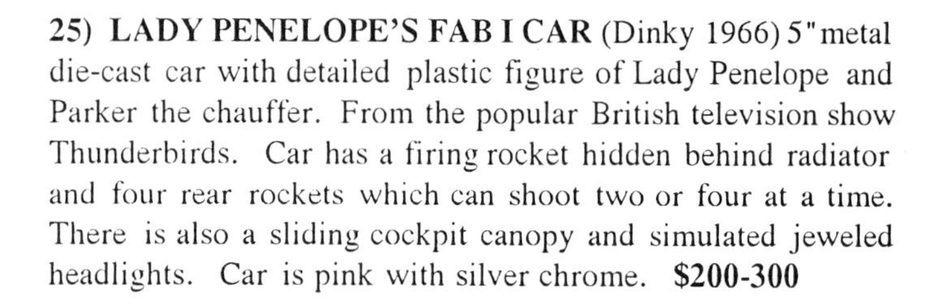

25) LADY PENELOPE'S FAB I CAR (Dinky 1966) 5"metal die-cast car with detailed plastic figure of Lady Penelope and Parker the chauffer. From the popular British television show Thunderbirds. Car has a firing rocket hidden behind radiator and four rear rockets which can shoot two or four at a time. There is also a sliding cockpit canopy and simulated jeweled headlights. Car is pink with silver chrome. **$200-300**

26

26) LADY PENELOPE'S FAB 1 CAR (J. Rosenthal Toys 1960'sBritish made) 10"long pink plastic friction drive replica of Lady Penelope's Fab 1 Car. Car is authentically detailed with clear plastic windshield dome, white tires, personalized license plates and silver chrome accents. Car has added feature of automatic ejecting guns which appear from front grill of car when hidden button is pressed. Car also has two detailed figures of Lady Penelope and her chauffeur, Parker. Box measures 11"x5"x3"deep. **$100-200**

GUMBY & POKEY

The Gumby Show debuted in 1957 and was hosted by Bobby Nicholson (of early "Claribel"fame from Howdy Doody). The tiny green clay figure was animated via a stop-motion technique similar in principal to RKO's 1933 film King Kong. Gumby's popularity ws revived in 1966 when new episodes were produced. Lakeside Toy Company headed the merchandising campaign.

1

2

1) **GUMBY DOT TO DOT BOOK** (Whitman 1968) 8"x11", 40-page dot-to-dot and coloring book. **$10-15**

2) **GUMBY ELECTRIC DRAWING SET** (Lakeside 1966) 10"x15"x4"deep box contains plastic light table/drawing stand, twelve black/white illustrated drawing sheets of Gumby and Pokey, an eraser-shaped Gumby, color pencils and sharpener. **$25-50**

3) **GUMBY HALLOWEEN COSTUME** (1968) Attractive unboxed costume of our favorite piece of clay. Body part is silk and glitter. **$25-40**

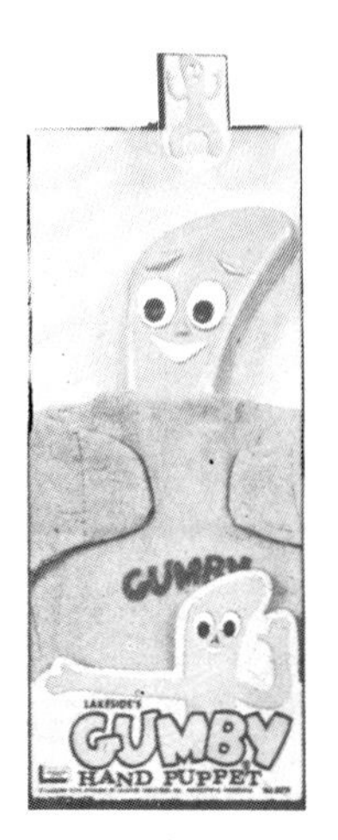

4

5

4) **GUMBY HANDPUPPET** (Lakeside 1965) 9" handpuppet with vinylhead and cloth body comes in poly bag with header card. **$15-30**

5) **GUMBY'S PAL POKEY HANDPUPPET** (Lakeside 1965) 8" orange handpuppet with vinyl plastic head and cloth body. Green lettering on chest reads "Gumby's Pal Pokey." **$25-40**

6

6) **GUMBY MODELING CLAY** (Chemtoy 1967) Boxed set comes with three tubs of modeling clay plus cut-out figures of Gumby and Pokey. **$25-50**

7

8

7) **GUMBY SUPER-FLEX FIGURE** (Lakeside 1965) 6"x9" display card holds 6" green rubber-like figure of Gumby which can be bent and twisted into a variety of poses. **$25-50**

8) GUMBY'S PAL POKEY SUPER-FLEX FIGURE (Lakeside 1965) 6"x9"display card holds 6" orange rubber-like figure of Pokey which can be bent and twisted into a variety of poses. **$25-50**

9

9) GUMBY ADVENTURE COSTUME STORE DISPLAY SIGN (Lakeside 1965) 12"x14"heavy cardboard display sign with easel stand-up features Gumby in the four adventure costumes produced by Lakeside for their 6" Gumby figure. **$50-75**

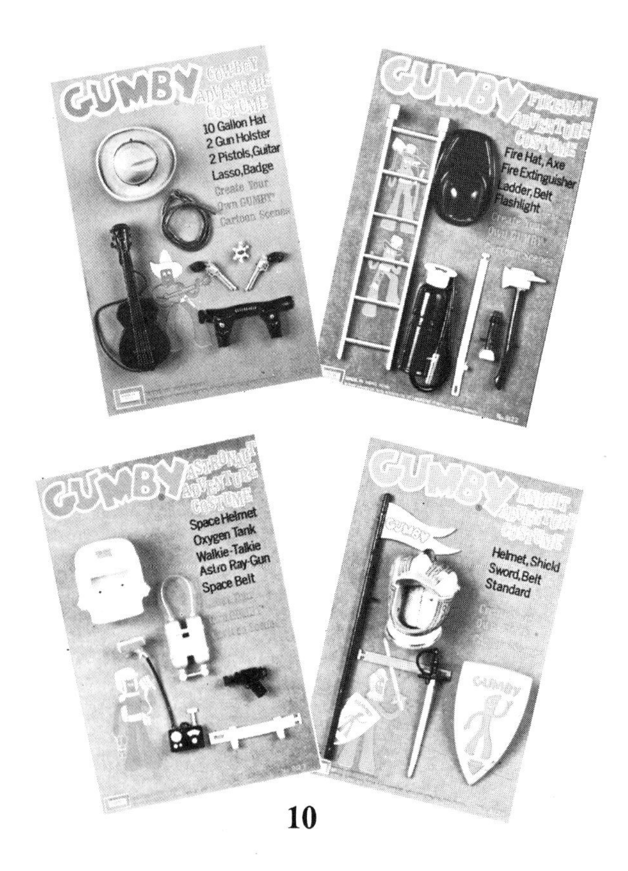

10

10) GUMBY ADVENTURE COSTUMES (Lakeside 1965) 6"x9"display card holds costume for the 6" Lakeside Gumby figure. Four different costumes were made, each came with a hat and several smaller accessories. (Astronaut, Fireman, Cowboy and Knight.) EACH: **$10-20**

11

11) GUMBY'S JEEP (Lakeside 1966) Yellow metal 8" long jeep with decals reading "Gumby's Jeep" on hood, both sides of jeep and seats. Fold-down windshield and rubber tires. Made for the 6" super-flex figures Lakeside also produced. Box is 10"x6"x6".**$50-75**

12) GUMBY TRAY PUZZLE (Whitman 1968) Inlay jigsaw puzzle in a 11"x14"frame tray. **$15-20**

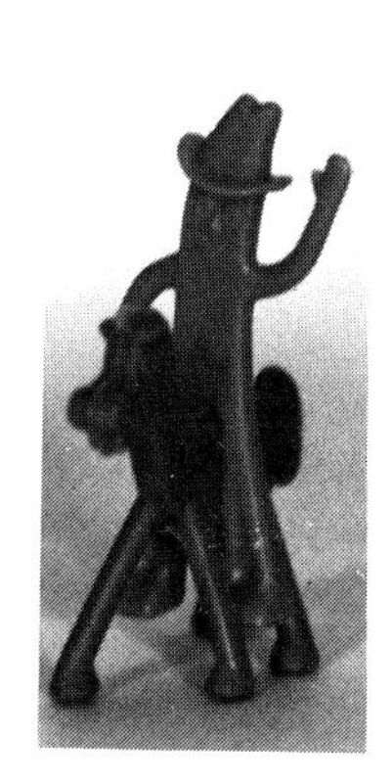

13

13) GUMBY & POKEY "PLAYFUL TRAILS" GAME (CO-5 Company 1968) 10"x20"box contains playing board and eight miniature figures of Gumby and Pokey (Gumby wearing a cowboy hat). Object of game is for Gumby to ride Pokey successfully into his corral without being "bucked off".**$45-60**

HANNA-BARBERA

In 1959, Bill Hanna and Joseph Barbera introduced the world to a new wve of cartoon characters and met with overwhelming success. The Flintstones, Huckleberry Hound, Yogi Bear and Quick Draw McGraw all became household names overnight--partiallydue to the enormous merchandising campaign that accompanied the project. Thousands of items were produced, from curtains to record players, and millions of units were sold. It was a new age for both television animation and toy manufacturers and neither industry has been the same since.

1

2

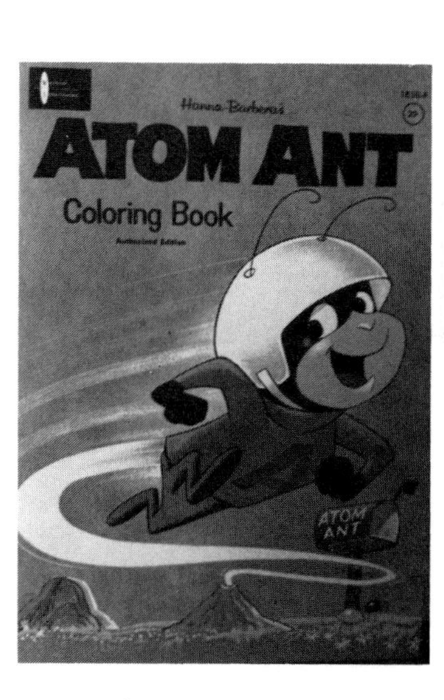

3

1) ATOM ANT & SECRET SQUIRREL LUNCHBOX WITH THERMOS (1966) Metal lunchbox features Secret Squirrel, Morocco Mole, Squiddly Diddly, and Chief Winchly on one side and Atom Ant on other. Thermos features above characters and Pa and Shag Rugg against a bright blue background. Box: **$100-150** Thermos: **$35-65**

2) ATOM ANT BUBBLE BATH CONTAINER (Purex 1965) 10" figural bubble bath container with hard plastic removable head. **$15-25**

3) ATOM ANT COLORING BOOK (Watkins-Strathmore 1965) 8"x10" hardback book contains reprinted Gold Key Comic Book stories. **$15-25**

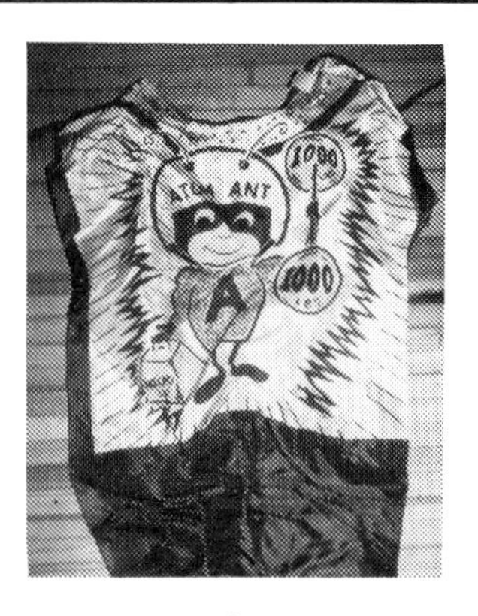

4

5

9

9) **ATOM ANT "SAVES THE DAY" GAME** (Transogram 1966) 10"x17"box. **$75-100**

4) **ATOM ANT HALLOWEEN COSTUME** (Ben Cooper 1965) 10"x12"window display box contains one-piece fabric bodysuit and mask. **$25-50**

5) **ATOM ANT JIGSAW PUZZLE** (Whitman 1965) 10"x12" box contains 100-piece jigsaw puzzle which assembles into a 14"x8" scene of Atom Ant at the computer in his secret headquarters. **$15-25**

10

11

6

8

10) **AUGIE DOGGIE TRAY PUZZLE** (Whitman 1960) Inlay jigsaw puzzle in a 11"x14"frame tray. **$10-15**

11) **AUGIE DOGGIE TRIVET** (Jellystone Park 1960's) Metal three-legged stand for setting under hot dishes/pans. Has painted embossed figure of Augie with a verse below. This was available only through gift shops of Jellystone Amusement parks in the 1960's. **$10-20**

6) **ATOM ANT MAGIC SLATE** (Watkins Strathmore 1967) 8"x12"illustrated cardboard display card holds lift-up erasable film sheet and comes with wood stylus. **$25-40**

7) **ATOM ANT PUSH UP PUPPET** (Kohner 1965) 3"plastic jointed mechanical puppet which moves about when bottom underneath base is depressed. **$15-20**

8) **ATOM ANT TRICKY TRAPEZE** (Kohner 1964) 3" plastic jointed mechanical figure is suspended from trapeze bar which is mounted to a base with a button on each side. When buttons are squeezed, the figure flips over the trapeze. **$20-25**

12

12) **AUGIE DOGGIE & LOPPY DE LOOP BOOK** (Whitman 1960) 5"x7"hardback book contains 80+ pages of comic book-style stories. Part of the Whitman Comic Book series. **$12-18**

13

16

13) **T-BONE DOG BISCUITS FEATURING AUGIE DOGGIE** (Gro-Pup 1962) 9"x7"box features Augie Doggie on front and back of box. **$50-100**

16) **THE BANANA SPLITS FIGURES** (Hasbro 1969) 4"x9" display card holds 4" soft painted rubber figure of a Banana Splits character. EACH: **$20-25**

14

14) **THE BANANA SPLITS "BANANA BUGGY" MODEL KIT** (Aurora 1969) 8"x5"x4"box contains all plastic assembly kit of a dune buggy and includes figures of the Banana Splits, decal sheet and instructions. **$100-125**

17

17) **THE BANANA SPLITS GAME** (Hasbro 1969) 20"x10" game contains playing pieces, game board and spinner. **$40-60**

15

15) **THE BANANA SPLITS CHALK BOARD** (Hasbro 1969) 24"x18"chalkboard mounted on stiff particle board. Comes with chalk, eraser and plastic pegs. **$25-50**

18

19

18) **THE BANANA SPLITS NUMBERED PENCIL COLOR SET** (Hasbro 1969) 11"x10"box contains twelve pre-numbered sketches and six color pencils. **$45-75**

19) **THE BANANA SPLITS PAINT BY NUMBER 'N FRAME SET** (Hasbro 1969) 14"x12"window display box contains large plastic frame, three pre-numbered sketches, six paint tablets and brush. **$65-100**

20

21

20) THE BANANA SPLITS "STAND-UP" RUB-ONS (Hasbro 1969) 11"x10"box contains ten die-cut cardboard figures with printed transfer sheets of various outfits which may be "rubbed-on" and transferred to the figure. Stylus included. **$50-75**

21) THE BANANA SPLITS STITCH-A-STORY SET (Hasbro 1969) 9"x9"box contains plastic frame, embroidery pictures, needle and thread. **$25-50**

22

23

22) THE BANANA SPLITS TALKING TELEPHONE (Hasbro 1969) 15"x8"x6"deep box contains plastic battery-powered pay-phone-style talking phone which operates by way of interchangeable records selected by phone dial. **$75-100**

23) BIRDMAN HALLOWEEN COSTUME (Ben Cooper 1967) 10"x12"window display box contains one-piece fabric bodysuit and mask. **$20-35**

24

24) FLINTSTONES BANK (1971) Barney & Bamm-Bamm 19" tall vinyl detailed bank. **$15-20**

25

26

25) FLINTSTONES BARNEY RUBBLE FIGURE NIGHTLIGHT (Snapit 1964) Color plastic 4" figure of Barney comes on 7"x4"card. **$12-18**

26) FLINTSTONES BIG GOLDEN BOOK (Golden Press 1960) 8"x10"hardback book contains 48-page story with full color story art throughout. **$15-25**

27

27) FLINTSTONES BOWLING PIN FIGURES (TransOgram 1960) Set of six hollow plastic 7"painted figures of Fred, Wilma, Barney, Betty, Dino and Baby Puss. EACH: **$8-12** SET: **$50-75**

28

28) FLINTSTONES "BRAKE-BALL" GAME (Whitman 1960) **$25-50**

29

29) FLINTSTONES CHALK BOX (Advance Crayon 1963) 4"x7"chalk box contains twelve color chalk sticks. **$15-20**

30

31

30) FLINTSTONES CIRCUS FIGURES (Kohner 1963) Large 10"x12"window display box contains ten different plastic 3" Flintstone characters which can be set up and re-arranged in a variety of different balancing positions. Set also includes balancing pole, trapeze, ladder, bar bell, rings, chairs and more. All parts are interchangable and snap together easily. **$50-75**

31) FLINTSTONES COLORING BOOK (Whitman 1960) 8"x11",80+ pages. **$10-15**

32) FLINTSTONES FRED VINYL FIGURE (Knickerbocker 1960) Colorful 15" figure. **$20-25**

33

33) FLINTSTONES FUZZY FELT (Standard Toykraft 1961) 10"x12" box contains colorful cloth/felt characters and accessory pieces which can be affixed to an illustrated background board. Same concept as Colorforms. **$25-50**

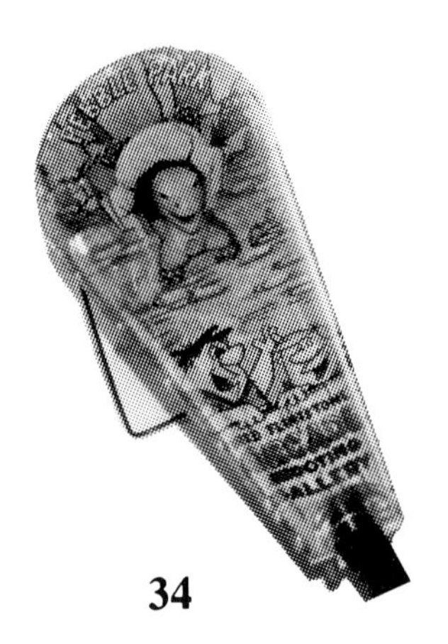

34

34) FLINTSTONES MECHANICAL SHOOTING GALLERY (Marx 1962) 14" tin litho target set with clear plastic dome. The main target is a plastic dinosaur head with open mouth. There is a 6" plastic pistol attached to the front of the target set that shoots small metal pellets. Comes in 24"x8"x8"box. **$200-250**

35

35) FLINTSTONES "MITT-FULL" GAME (Whitman 1962) **$25-50**

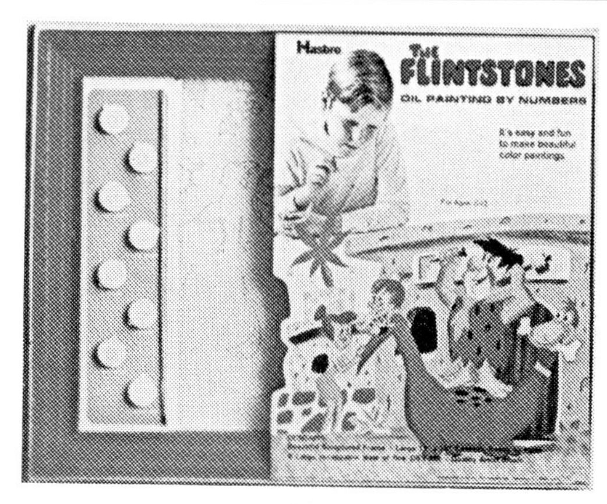

38

37) **FLINTSTONES OIL PAINTING BY NUMBERS SET** (Hasbro 1966) 26"x18"box includes same-size plastic frame, a pre-numbered, pre-sketched canvas board, eight vials of oil paints and brush. **$40-70**

39

39) **FLINTSTONES PILLOW CASE** (1960) Cotton cloth pillow case with printed image of Fred at piano. **$10-20**

36

36) **FLINTSTONES MOTORIZED MODEL KITS** (Remco 1961) 20"x12"x4"box contains all plastic assembly kit and includes small battery operated electric motor which, when assembled, powers the toy in a forward direction. Kits contain cardboard stand-up figures.

A)	Paddy Wagon	**$150-250**
B)	Yacht	**$100-200**
C)	Car and trailer	**$100-200**

40

38) **FLINTSTONES PARTY PLACE SET** (Reed 1969) Nice four piece set includes tablecloth, 24 napkins, 8 small and 8 large plates. **$20-25**

40) **FLINTSTONES PLAY SET** (Marx 1962) 24"x15"x5" deep box contains plastic play set which includes 2.5"figures of the Flintstones (Fred, Wilma, Barney, Betty, Dino) plus other Bedrock civilians, cars, houses and many accessories including instructions and plastic layout sheet. (Note: This playset was recently re-issued by a company in Florida and will say so on the side of the box.) **$300-600**

41

41) **FLINTSTONES PRE-HISTORIC ANIMAL RUMMY PLAYING CARDS** (Ed-U-Card 1960) 3.5"x3.5"box contains color playing cards of Flintstone characters and dinosaurs. **$10-20**

42

42) **FLINTSTONES "STONE AGE" BOARD GAME** (TransOgram 1961) **$20-35**

43

43) **FLINTSTONES TILE SQUARE PUZZLE** (Roalex 1962) 3"x3"black plastic frame contains 15 illustrated square tiles and one empty tile space. Object is to rearrange tiles to form correct illustrations of Fred, Wilma, Barney and Betty. Comes on illustrated 5"x8"card. **$25-30**

44

44) **FLINTSTONES-WELCH'S STORE DISPLAY POSTER** (Welch's 1962) 18"x22"color poster features all the Flintstone characters promoting Welch's Fruit Drinks. **$50**

45 46

45) **DINO PUSH BUTTON PUPPET** (Kohner 1964) 3" painted plastic jointed mechanical figure puppet which moves in a variety of poses when bottom of base is depressed. **$10-20**

46) **DINO THE DINOSAUR BENDY FIGURE** (Bendy 1961) 10" salmon color foam bendable figure of Dino with painted accents on face and polka dot body. Arms, tail and neck bend. **$25-75**

47

47) **DINO THE DINOSAUR TO COLOR BOOK** (Whitman 1961) 8"x11",80-pages. **$15-20**

48

48) DINO THE DINOSAUR GAME (Transogram 1961) 10"x17"box. **$35-50**

49

49) FLINTSTONES DINO THE DINOSAUR TIN WIND-UP (Marx 1961) Purple tin litho Dino stands 6" tall and 9" long. There is an on/off switch on his back which, after toy is wound up and switch is activated, will walk forward as his mouth opens and shuts. His name "Dino the Dinosaur" is printed on both sides of his tail. **$200-300**

50

50) FRED FLINTSTONE ON DINO BATTERY TOY (Marx 1962) Large color 21"x12"battery-operated Dino has hard vinyl head and soft plush purple fabric body. There is a 3" tin figure of Fred with detailed vinyl head and cloth outfit and tie seated in a tin pagoda-style crane operators carriage and he is holding onto a reign attached to Dino. Toy, when activated, moves on hidden wheels, and when it bumps into something, will back up and turn as its tail wiggles. Dino's neck moves up and down and right to left as his mouth opens and shuts. Fred rocks back and forth creating the impression he is pulling on the reigns and "operates" Dino's motions. Box measures 13"x16"x8"and features colorful art on lid and all four side panels. **$500-800**

51

52

51) FRED FLINTSTONE AND DINO PUSH BUTTON PUPPET SET (Kohner 1964) 7"x5"x2"window display box contains 3"painted plastic jointed mechanical puppets of Fred and Dino and are each mounted on a base. There are holes on the back of the box where puppets can be operated without removing them from box. When bottom of base is depressed, figures move about. **$45-75**

52) FRED FLINTSTONE PUSH BUTTON PUPPET (Kohner 1964) 3" painted, plastic, jointed, mechanical figure puppet which moves in a variety of poses when bottom of base is depressed. **$10-20**

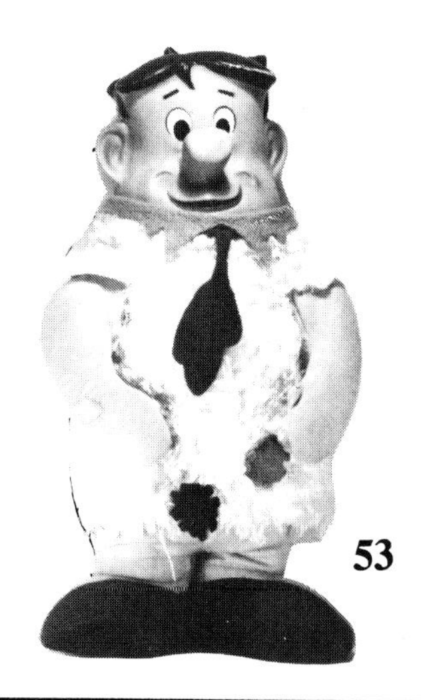

53

54

53) **FRED FLINTSTONE PLUSH DOLL** (Knickerbocker 1960) 18" plush doll with vinyl plastic head. **$25-50**

54) **HOPPY HOPPAROO PUSH BUTTON PUPPET** (Kohner 1964) 3" painted, plastic, jointed, mechanical figure puppet which moves in a variety of poses when bottom of base is depressed. **$15-20**

55

55) **FLINTSTONE'S HOPPY THE HOPPAROO GAME** (Transogram 1965) 10"x20"box. **$50-75**

56

56) **BAMM-BAMM BUBBLE PIPE** (Transogram 1963) 5"x7" display card holds plastic pipe with Bamm-Bamm's head forming the front of the pipe. **$12-15**

57) **BAMM-BAMM "COLOR ME HAPPY" GAME** (Transogram 1963) 10"x17"box. **$35-60**

58 **59**

58) **BAMM-BAMM DOLL** (Ideal 1963) 16" tall vinyl plastic posable doll dressed in leopard skin suit which buttons with bone; hat and club. Comes in 14"x12"x5"deep window display box. Inside of box is designed like his room. This doll came in two sizes:

Large 16" Bamm-Bamm	**$150-200**
Small 12" Tiny Bamm-Bamm	**$100-150**

59) **BAMM-BAMM PUSH BUTTON PUPPET** (Kohner 1964) 3" painted plastic jointed mechanical figure puppet which moves in a variety of poses when bottom of base is depressed. **$10-20**

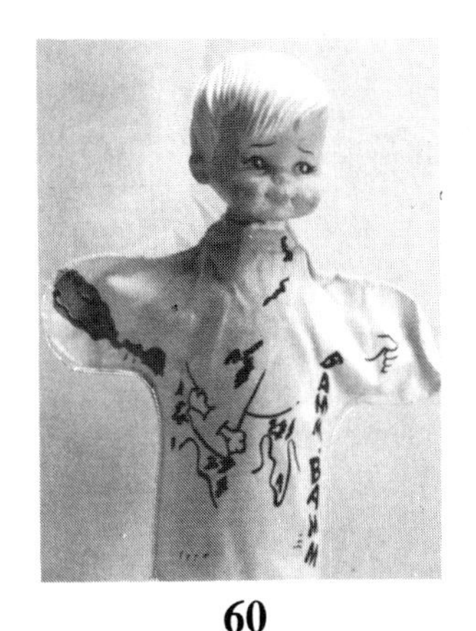

60

61

60) **BAMM-BAMM HANDPUPPET** (Ideal 1963) 12" handpuppet with illustrated cloth body and soft vinyl plastic head. **$20-30**

61) **BAMM-BAMM MUG** (Flintston Vitamins 1972) 3" tall hard plastic mug molded in the image of Bamm-Bamm's head. This mug originally contained Flintstone vitamins and the mug was a premium. **$5-10**

62

62) **PEBBLES BUBBLE BATH** (Roclar 1963) 4"x10"x2"deep box with flip-up display lid and 3-D pop-up figures of Fred and Wilma bathing Pebbles. Box contains 24 packets of powder bubble bath soap in four different colors. There are four different packet illustrations of Pebbles. **$50-75**

63

64

63) **"BABY" PEBBLES DOLL** (Ideal 1962) 16" tall plastic posable doll with saran hair upswept into a ponytail and kept in place by plastic bone. Doll is dressed in two-piece outfit featuring leopard skin top with matching robe and blanket. Box is 9"x16"x5"deep. **$200-250**

64) **"TINY" PEBBLES DOLL** (Ideal 1964) 12" vinyl plastic rendition of the 16" Pebbles doll. Comes in 12"x14"window display box. The inside of box is designed to look like her room. **$150-200**

65

65) **PEBBLES AND HER CRADLE** (Ideal 1964) Large 18"x16"x8"deep window display box contains 12" Pebbles doll and plastic rocking log cradle. **$250-325**

66

66) **PEBBLES FLINTSTONE CRAYON BY NUMBER AND STENCIL SET** (TransOgram 1963) 9"x16" window display box contains cut-out stencils of Pebbles, eight pre-numbered sketches to color, 24 crayons and sharpener. **$35-50**

67

67) **PEBBLES FLINTSTONE GAME** (Transogram 1962) 10"x17"box. **$35-65**

68

68) **PEBBLES FLINTSTONE MAGNETIC FISH POND GAME** (TransOgram 1963) 16"x24" window display box contains large vacuform plastic fish pond, two miniature plastic fishing rods with working reels and magnetic hooks and nine magnetic plastic fish that stand or float if water is added to pond. **$50-75**

69

70

69) **PEBBLES FLINTSTONE SEWING CARDS** (TransOgram 1963) 9"x17" box contains several oversized illustrated cards with small holes along the outline of the character(s) which can be sewn along and "traced" by plastic-tipped yarn laces (which are included). **$25-35**

70) **PEBBLES AND BAMM-BAMM PAPER DOLLS** (Whitman 1965) Pebbles and Bamm-Bamm. **$15-25**

71

71) **PEBBLES FLINTSTONE SLIDE TILE PUZZLE** (Roalex 1963) 8"x5"display card holds 3"x3"b/w plastic puzzle with movable tiles to form image of Pebbles. **$20-30**

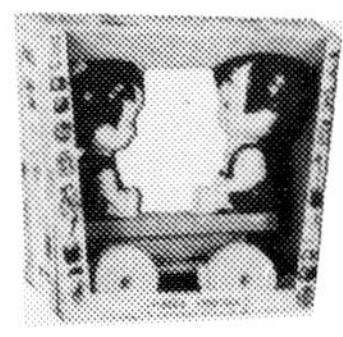

72

72) **PEBBLES AND BAMM-BAMM PULL TOY** (TransOgram 1963) 12"x11"x5"window display box contains 6" figures of Bamm-Bamm and Pebbles sitting on a stone age cart. When toy is pulled, figures bob up and down in a "see-saw" action. **$50-75**

73

73) **PEBBLES AND BAMM-BAMM CRAYON BY NUMBER AND STENCIL SET** (TransOgram 1963) 9"x16" window display box contains cut-out stencils of Pebbles and Bamm-Bamm, eight pre-numbered sketches to color, 24 crayons and sharpener. **$35-50**

74

74) **PEBBLES AND BAMM-BAMM PAINT BY NUMBER COLORING SET** (TransOgram 1965) 16"x9"window display box contains plastic paint tray with eight inlaid paint tablets, brush and eight pre-numbered sketches. **$40-60**

75

75) PEBBLES AND BAMM-BAMM PENCIL BY NUMBER COLORING SET (TransOgram 1965) 16"x9" window display box contains eight pre-numbered sketches, six color pencils and sharpener. **$35-65**

76

76) PEBBLES FLINTSTONE PULL-TOY (TransOgram 1963) 12"x11"x5" deep window display box contains 10" tall plastic figure of Wilma pushing Pebbles in a turtle shell baby carriage. When toy is pulled, Pebbles bounces up and down. **$50-75**

77 **78**

79

77) WILMA FLINTSTONE PUSH BUTTON PUPPET (Kohner 1964) 3" painted plastic jointed mechanical figure puppet which moves in a variety of poses when bottom of base is depressed. **$10-20**

78) PEBBLES FLINTSTONE PUSH BUTTON PUPPET (Kohner 1964) 3" painted plastic jointed mechanical figure puppet which moves in a variety of poses when bottom of base is depressed. **$10-20**

79) FLINTSTONES WILMA BENDABLE FIGURE (Giant Plastic 1968) 4" figure of Wilma on color card. **$15-20**

80 **81**

80) "SOUNDS OF THE FLINTSTONES" CHILD'S RECORD (Golden 1961) Contains original theme song and original TV voices. Attractive picture sleeve. **$8-12**

81) ROCKY THE BATTERY OPERATED CAVEMAN (Marx 1962) Although not a licensed toy, the features of Rocky are a dead-ringer to Fred Flintstone. Toy is designed to roll across floor, and is 4" tall and has soft vinyl head and colorful tin litho body. **$25-50**

82

82) FUNKY PHANTOM BOARD GAME (MB 1971) **$10-15**

83

83) HANNA-BARBERA BATHTIME BUBBLE FUN (Milvern 1965) 4"x10"x2" deep window display box contains 20 packets of soap powder, each with a different illustration of Hanna-Barbera characters. **$25-50**

84) HANNA-BARBERA CEREAL BOXES. During the 1960's Kelloggs used many of the Hanna-Barbera characters to promote their cereal. Below we have listed a few of the boxes.

85

86

85) ALL STARS (Kelloggs 1963) Canadian distributed boxes of All Star cereal featured Huckleberry Hound. **$150-250**

86) COCO POPS (Kelloggs 1963) Canadian/British distributed boxes of Coco Pops feature Mr. Jinx, Pixie and Dixie. **$200-300**

87

88

87) COCOA KRISPIES CEREAL (Kelloggs 1964) Hanna-Barbara's Snagglepuss was Kellogg's official mascot for Cocoa Krispies for a short time in the early Sixties and was prominently featured on several boxes. Our photo example is from a 1964 box with cut-out antique car trading cards on back. **$250-400**

88) CORN FLAKES CEREAL BOX (Kelloggs 1960) Limited edition box commemorating one year birthday of Yogi Bear. There is a free Yogi Bear birthday comic book premium offered inside. **$400-500**

89

90

89) HONEY SMACKS (Kelloggs 1969) The Banana Splits were on the boxes of Kelloggs Sugar Smacks from 1968-70. Our photo shows a 1969 box featuring Bingo on the front and toy train engine premium on back. **$100-200**

90) OK's CEREAL BOX (Kelloggs 1960) "The Best in Oats" came in the shape of little O's and K's and featured Yogi Bear on the box as the official mascot. **$200-300**

91

91) SUGAR SMACKS CEREAL BOX (Kelloggs 1963) Hanna-Barbera's Quick Draw McGraw was the official Sugar Smacks mascot from 1963-64. Our photo shows a 1963 box. **$150-250**

92

92) VARIETY PACK-SIZE BOXES (Kelloggs 1960-69) Smaller 3"x4"variety pack-size boxes featuring Hanna-Barbera characters usually have a value of 50-60% of the larger size boxes. Our photo example shows Sugar Stars with Huckleberry Hound and Sugar Smacks with Quick Draw McGraw. EACH: **$50-100**

93

93) KELLOGGS VARIETY PACK TRAY BOTTOMS (Kelloggs 1960's) Most all of the Kelloggs variety packs which contained children's sweetened cereal featuring Hanna-Barbera characters often had an activity such as cut-outs, puzzles, etc. on the bottom of the tray. Our photo example shows a Huckleberry Hound cut-out picture puzzle. **$20-35**

94

94) KELLOGGS-HANNA-BARBERA STORE DISPLAY POSTER (Kelloggs 1964) 35"x24" glossy full color store display poster features Huck Hound, Quick Draw McGraw, and Snagglepuss playing basketball with Kellogg characters Tony the Tiger and Sugar Pops Pete. Each is tossing a box of their own cereal with their pictures on front. **$250-500**

95

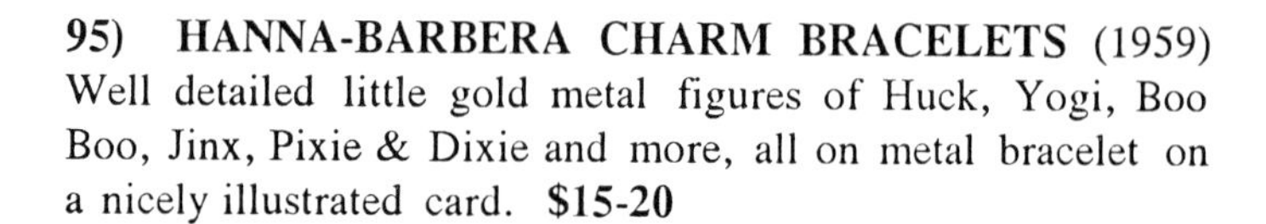

95) HANNA-BARBERA CHARM BRACELETS (1959) Well detailed little gold metal figures of Huck, Yogi, Boo Boo, Jinx, Pixie & Dixie and more, all on metal bracelet on a nicely illustrated card. **$15-20**

96

96) HANNA-BARBERA GIVE-A-SHOW PROJECTOR SET (Kenner 1963) 11"x18" box contains plastic battery-powered slide-film projector and 16 color slide strips. Kenner made a variety of Hanna-Barbera sets, each with a predominant illustration of a certain character on the box lid (Jetsons, Jonny Quest, Huck Hound, etc.). Our photo shows the Flintstones set. **$25-50**

97

97) HANNA-BARBERA "SNAP-TOGETHER" FIGURES (Kelloggs 1960) Plastic three-piece 2" figures of Huckleberry Hound, Yogi Bear, Boo-Boo and Jinx the Cat which detached from the head and feet. These figures came free inside Corn Flakes and Sugar Frosted Flakes and there is also a figure of Tony the Tiger. EACH: **$20-40**

98

98) HANNA-BARBERA STAMP SET (1963) Large deluxe stamp set contains many early and obscure H-B characters like Snagglepuss, Snooper & Blabbermouth, Top Cat & gang and the Flintstones. Twenty stamps in all. Box is 14"x10". **$35-65**

99

99) HANNA-BARBERA TV-TINY KINS (Marx 1962) 13"x10"window display box contains 34 detailed painted figure from five different H-B cartoon shows. Each show has six to seven figures mounted on a 10"long card with scene backdrop and are about 2" tall. Shows include: Flintstones, Quick Draw McGraw, Huckleberry Hound, Top Cat and Yogi Bear. **$300-400**

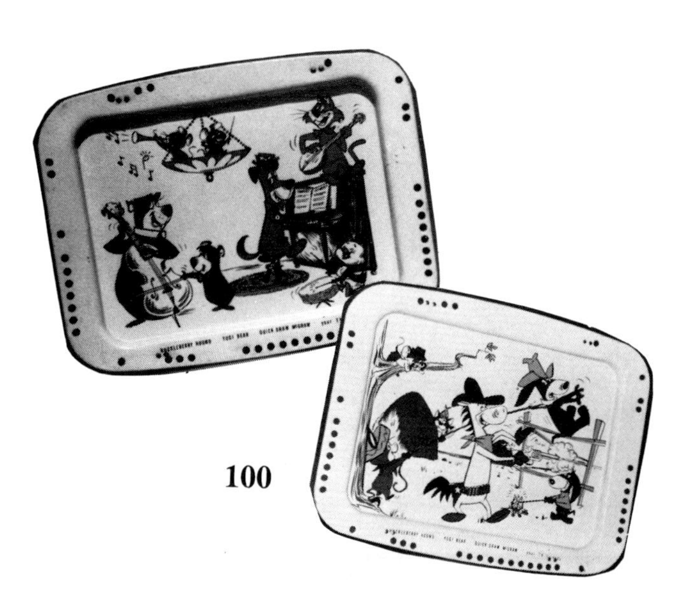

100

100) HANNA-BARBERA TV TRAYS (1959) Sturdy 16"x12" tin tray with painted scene of Huck, Yogi, and Quick Draw McGraw. EACH: **$20-35**

101

101) HANNA-BARBERA TV CARD GAME (Ed-U-Cards 1961) Large 14"x4"deluxe boxed set contains several different card games. All cards are colorfully illustrated and many different characters featured. **$25-40**

102

102) HANNA-BARBERA XYLOPHONE (1960) Colorful little litho instrument with each key featuring a different H-B character. **$20-35**

103

103) HUCKLEBERRY HOUND "BUMPS" BOARD GAME (TransOgram 1960) **$25-40**

104

105

104) HUCKLEBERRY HOUND BUTTON (1960) 3" diameter color button with yellow background reads "Huckleberry Hound for President". **$8-15**

105) HUCKLEBERRY HOUND FIGURAL BANK (Knickerbocker 1960) 11" hard plastic bank with white/black painted accents depict Huck in top hat and tuxedo coat. **$15-25**

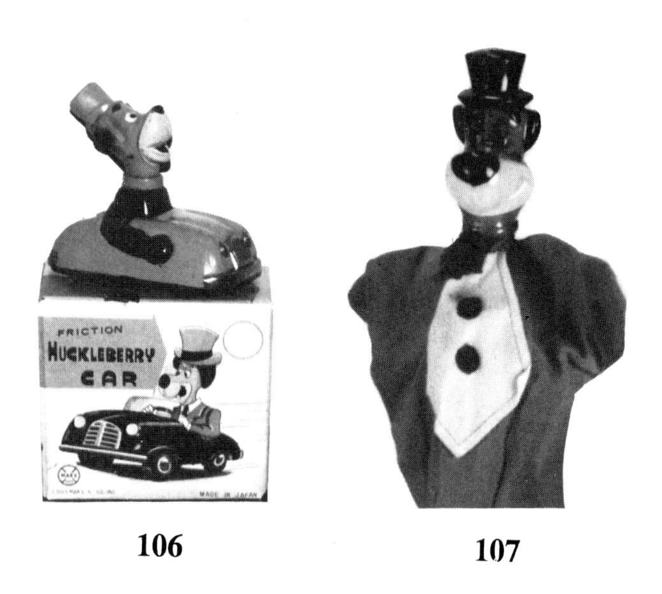

106 **107**

106) HUCKLEBERRY HOUND CAR (Marx 1960) 5" long tin convertible car with tin half-body and vinyl head of Huck Hound. Car is friction drive. Box is 5"x5"x4".**$50-100**

107) HUCKLEBERRY HOUND HAND PUPPET (Kelloggs 1960) 9" puppet with hard plastic head and cloth body made of red, white and blue fabric. Offered exclusively as a mail order premium through Kelloggs Corn Flakes. **$25-50**

108

108) HUCKLEBERRY HOUND PENCIL CASE (Hasbro 1960) 4"x8"x1½" thick cardboard box with snap-open lid and pull-out drawer. Illustrated colored paper decal features Huck, Yogi, Boo-Boo, Mr. Jinx and Pixie and Dixie on a merry-go-round. **$15-25**

109

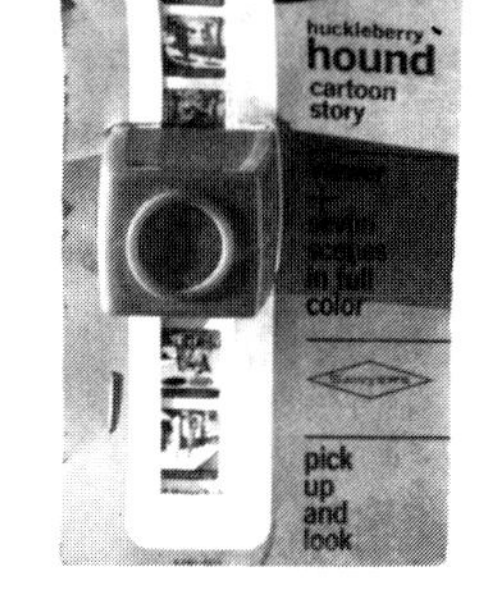

110

109) HUCKLEBERRY HOUND PUSH BUTTON PUPPET (Kohner 1964) 3" painted plastic jointed mechanical figure puppet which moves in a variety of poses when bottom of base is depressed. **$10-20**

110) HUCKLEBERRY HOUND STORY VIEWER AND FILM SET (Sawyer 1960) Small 2"x1"x1½" plastic viewer and long rectangular film strip come attached on 4"x8" display card. **$12-15**

111

111) HUCKLEBERRY HOUND TELLS STORIES OF UNCLE REMUS RECORD ALBUM (HBR 1963) 33-1/3 rpm record contains songs and stories.

- **A)** original **$15-20**
- **B)** 1977 re-issue **$3-5**

112

112) HUCKLEBERRY HOUND TRAY PUZZLE (Whitman 1960) With Pixie and Dixie. Inlay jigaw puzzle in 11"x14" frame tray. **$12-15**

113

113) HUCKLEBERRY HOUND TV WIGGLE BLOCKS (Kohner 1962) 14"x9"window display box contains ten small 2"x2"x1"plastic TV sets with flicker-flasher images of Huck and friends in the picture screen. **$75-125**

114

114) HUCKLEBERRY HOUND WESTERN BOARD GAME (MB 1959) **$20-40**

115

115) HUCKLEBERRY HOUND & HIS FRIENDS CHILD'S RECORD (Golden 1959) **$8-12**

116

116) HUCKLEBERRY HOUND AND FRIENDS COLORFORMS SET (1960) 8"x12"box contains die-cut thin vinyl stick-on character pieces and accessory pieces, plus background scene for placement of characters, and fold-out instruction booklet. **$25-40**

117

117) HUCKLEBERRY HOUND AND HIS FRIENDS GIANT GOLDEN BOOK (Golden 1965) Large 9"x11" hardback book with 80+ page full color art stories including Touche Turtle and Ruff and Reddy. **$20-30**

118

118) HUCKLEBERRY HOUND AND FRIENDS LUNCHBOX WITH THERMOS (1961) BOX: **$75-100**
THERMOS: **$20-35**

119) **HUCKLEBERRY HOUND AND FRIENDS WASTE PAPER BASKET** (1960) Child's tin litho basket. **$35-65**

122) **HUCKLEBERRY HOUND AND YOGI BEAR BIG GOLDEN BOOK** (Golden Press 1960) 8"x10"hardback book contains 48-page story with full color story art throughout. **$15-25**

120

120) **HUCKLEBERRY HOUND AND QUICK DRAW WALL PLAQUES** (1960) 11" figural wall hanging plaques molded in hard plastic and colorfully painted. **EACH: $10-20**

123

123) **HUCKLEBERRY HOUND AND YOGI CEREAL BOWL AND MUG SET** (Kelloggs 1961) 6" diameter thick, hard plastic bowl features three raised images of Huck Hound on the side of the bowl, painted in blue, and depicting him as a jet pilot, sheriff, and floating on a raft. The 4" tall mug is molded in the shape of Yogi's head with color accents of brown, flesh, green, black and white. **$25-50**

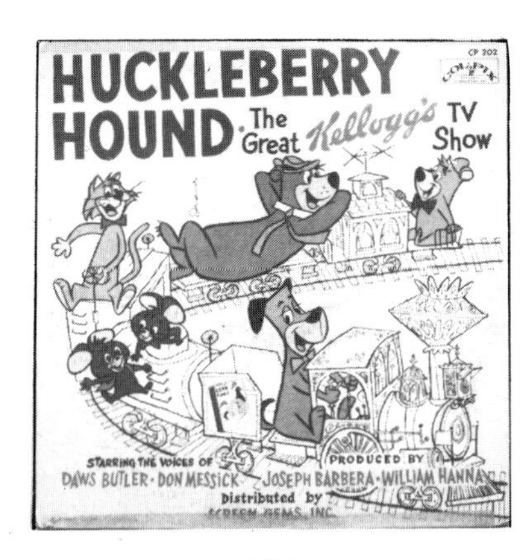

121

121) **HUCKLEBERRY HOUND & THE GREAT KELLOGG'S TV SHOW RECORD ALBUM** (Colpix 1960) 33-1/3 rpm record album originally available through specially marked boxes of Kelloggs cereals. **$20-30**

124) HUCK HOUND AND YOGI BEAR CEREAL SPOONS (Kelloggs 1960) 5" stainless steel spoon with die-cut detailed image of Huck or Yogi at the end of the handle. Their names embossed down the length of the handle. Available through Kelloggs' Corn Flakes and Shredded Wheat as a mail order premium. EACH: **$8-15**

125

125) HUCK & YOGI FIGURAL NIGHTLIGHT (1961) 7" tall molded soft vinyl figures of Huck and Yogi standing at opposite sides of a television set with embossed facial image of Mr. Jinx on the screen. Light turns on by switch in back and lights up the television screen. **$50-75**

126

127

126) HUCKLEBERRY HOUND & YOGI BEAR TALKING MOVIE WHEELS (Movie Wheels, Inc. 1960) Colorful 11"x12"display card contains rotating "movie wheel" scenes which are manually played by a dial. Also included is a 7" record which narrates a story based on the scenes. Display card is double-sided and has story/movie wheel scenes for Huck and Yogi. **$12-18**

127) INCH HIGH PRIVATE EYE GUNS (1972) 5"x6"card contains two miniature guns. **$8-10**

128 **129**

128) IMPOSSIBLES MAGIC SLATE (Watkins-Strathmore 1969) 8"x12"illustrated cardboard display card holds lift-up erasable film sheet and comes with wood stylus. **$15-25**

129) IMPOSSIBLES HALLOWEEN COSTUME (Ben Cooper 1967) 10"x12"box contains mask and one-piece fabric bodysuit with illustration of Impossibles on front. **$25-50**

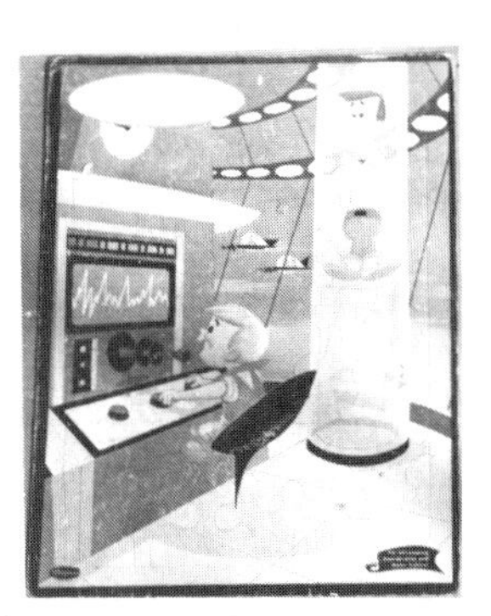

130 **131**

130) JETSONS "ASTRO" WINDUP HOPPER (Marx 1963) 4" tall grey tin litho figure of Astro with hard vinyl brown ears, grey plastic antenna tail and accented face highlights. When toy is wound up, body bobs up and down and legs "hop". **$200-300**

131) JETSONS FRAME TRAY PUZZLE (Whitman 1962) Inlay jigsaw puzzle in a 11"x14"frame tray of George and Elroy. **$25-35**

132

132) JETSONS FUN PAD GAME (MB 1963) Colorful boxed game comes with several small spacecraft with little figures riding inside. The object is to place as many spacecraft on the space pad as possible without tipping the balance. **$75-100**

133

133) JETSONS JIGSAW PUZZLE (MB 1963) 8"x10"box contains 70-piece jigsaw puzzle which assembles into a 14"x18" scene of space traffic cop chasing George Jetson. **$25-50**

134

134) JETSONS LUNCHBOX (Aladdin 1963) Metal blue dome top box with space graphics of Jetsons. Comes with steel thermos. Box: **$500-700** Thermos: **$75-100**

135

135) JETSONS MAGIC SLATES (Watkins/Strathmore 1963) 8"x14"diecut cardboard with magic erasable slate and pencil. One features family in spaceship; one features Elroy in space scooter with family around him (see photos). EACH: **$35-50**

136

136) JETSONS "OUT OF THIS WORLD" BOARD GAME (TransOgram 1963) 10"x20"box contains playing board, four miniature plastic space cards, two plastic comets, spinner disc and die-cut cardboard stand-up figures of George, Jane, Astro, Judy, Elroy and Rosie the Robot. **$100-150**

137) JETSONS "ROSIE THE ROBOT" GAME (Transogram 1963) 9"x17"box contains playing board, four cardboard character figures, spinner disc and 16 cardboard playing pieces. **$100-150**

138

138) JETSONS "ROSIE THE ROBOT" HALLOWEEN COSTUME (Ben Cooper 1963) 10"x12"window display box contains polyester body royal blue dress showing Rosie. Bottom of dress shows the heads of all the Jetsons and silver glitter is designed around dress to add balance. **$75-100**

139

139) JETSONS "HOPPING ROSIE" WIND-UP FIGURE (Marx 1962) 4" tall blue/black tin litho figure of Rosie with plastic side antennas. When toy is wound up, its body bobs up and down and legs "hop". **$250-350**

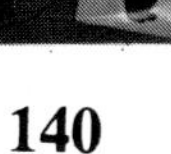

140

141

140) JETSONS "3 ON 1" RECORD SET (Golden 1963) 7"x8" cardboard sleeve shows three other Jetson's record sleeves including "George & Jane Jetson," "Rosey the Robot & Astro," and "Judy & Elroy." The record contains these three recordings. **$20-40**

141) SONGS OF THE JETSONS CHILDREN'S RECORD (Little Golden Records 1963) 7"x8"stiff paper sleeve contains 78 rpm record featuring Jane and Judy Jetson singing "Push Button Blue" and "Rama Rama Zoom." **$20-30**

142

142) JONNY QUEST CARD GAME (MB 1963) 6"x10"box contains illustrated playing cards, score pad, and vacuform playing card tray. Box lid features Jonny Quest, "Race", Bannon, and Dr. Benton Quest. **$50-100**

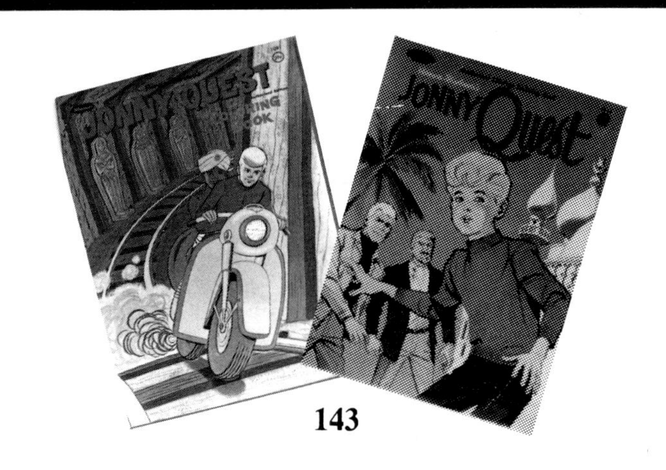

143

143) JONNY QUEST COLORING BOOKS (Whitman 1965) EACH: **$25-30**

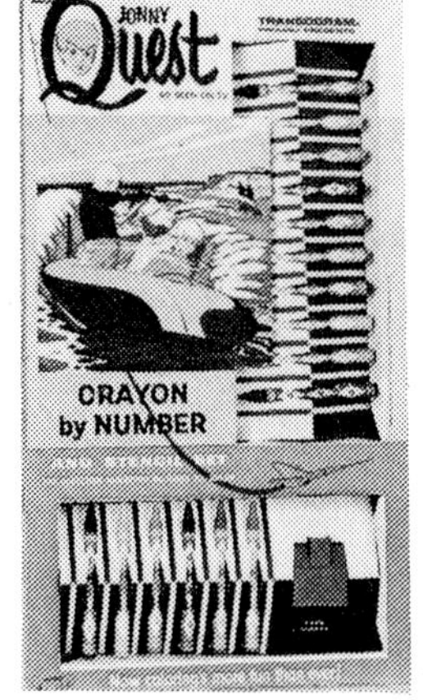

144

144) JONNY QUEST CRAYON BY NUMBER AND STENCIL SET (TrasOgram 1965) 9"x16"window display box contains cut-out stencils of Jonny Quest characters, paper, six crayons and sharpener. **$75-125**

145) JONNY QUEST CRAYON BY NUMBER SET (Transogram 1965) 17"x10"window display box contains six pre-numbered sketches and 16 crayons. **$100-125**

146

146) JONNY QUEST GAME (Transogram 1964) 10"x20" box. **$100-150**

147

147) JONNY QUEST PAINT BY NUMBER COLORING SET (TransOgram 1965) 16"x9"window display box contains plastic paint tray with eight inlaid paint tablets, brush and eight pre-numbered sketches. **$100-125**

148) JONNY QUEST PAINT BY NUMBER SET (Transogram 1965) 10"x17"window display box contains six pre-numbered sketches and six colored pencils. **$100-125**

149

149) JONNY QUEST PENCIL BY NUMBER COLORING SET (TransOgram 1965) 16"x9"window display box contains eight pre-numbered sketches, six color pencils and sharpener. **$75-125**

150

151

150) JONNY QUEST TV ANNUAL (British publisher 1965) 8"x10"hardback book contains reprinted Gold Key Comic Book stories. **$15-25**

151) LIPPY THE LION BUBBLE BATH CONTAINER (Purex 1964) 11"plastic figural soap container with removable head. **$20-30**

152

152) LIPPY THE LION GAME (Transogram 1962) 10"x17" box, **$50-100**

153

153) MAGILLA GORILLA AND FRIENDS RING TOSS 'N' BOWL GAME (Ideal 1964) 15"x19"window display box contains 5" soft hollow plastic figures of Magilla Gorilla, Punkin' Puss, Mushmouse, Ricochet Rabbit and Droop-Along Coyote which serve as bowling pins or target stakes for a ring toss game. Two bowling balls and two ring hoops included. **$50-100**

154

154) **MAGILLA GORILLA CANNON** (Ideal 1964) Large 7"x11"x12"window display box contains large plastic wheeled cannon with Magilla in driver's seat. Cannon is spring-loaded and comes with cannon balls. Cannonballs are loaded into barrel until spring release is caught. When the toy is pushed, spring releases firing the cannon ball. **$100-200**

155

156

155) **MAGILLA GORILLA FIGURE PULL TOY** (Ideal 1964) Large 10"plastic figure of Magilla in a box crate. **$25-35**

156) **MAGILLA GORILLA HAND PUPPET** (Knickerbocker 1965) 10"handpuppet with molded vinylhead and fabric body. **$20-30**

157

157) **MAGILLA GORILLA GAME** (Ideal 1964) 10"x20"box. **$50-100**

158

158) **MAGILLA GORILLA PLASTIC STICK-ONS** (Standard Toykraft 1964) Similar to a colorforms set. Thin vinyl plastic body parts of characters can be arranged on 8"x8.5"illustrated board. Comes in 9x13"window display box. **$25-50**

159

160

159) **MAGILLA GORILLA "POP-OUT" TARGET BARREL GAME** (Ideal 1964) 5" tall brightly colored plastic can features an extended target of Magilla's portrait which, when hit, releases a soft molded vinylhead of Magilla. Plastic spring loaded gun and three darts included. Box is 8"x8"x6". **$45-85**

160) **MAGILLA GORILLA "TWISTABLES"** (Ideal 1964) 8"posable stuffed doll with soft vinylface. Comes in 10"x5"x4" deep window display box. **$35-50**

161 **162**

161) **MAGILLA GORILLA PULL TOY** (Ideal 1964) 6" painted soft vinyl squeeze figure of Magilla mounted on yellow plastic wagon with pull string. Originally packaged in polybag with illustrated header card. **$20-30**

162) **MAGILLA GORILLA PUSH BUTTON PUPPET** (Kohner 1964) 3" painted plastic jointed mechanical figure puppet which moves in a variety of poses when bottom of base is depressed. **$15-20**

166

166) MUSHMOUSE AND PUNKIN PUSS TARGET GAME (Ideal 1964) Game consists of 17"x14" stand-up die-cut plastic target of a mountain shack with die-cut cardboard figures of Punkin' Puss and Mushmouse. When hit by a rubber-tipped dart, figures disappear and pop up in a new spot. Spring loaded gun and three safety darts included. Box is 18"x15"x4" deep. **$75-125**

163

163) **MUSHMOUSE & PUNPKIN PUSS GAME** (Ideal 1964) 10"x20" box. **$35-75**

167

167) MUSHMOUSE PULL TOY (Ideal 1964) 6" painted soft vinyl squeeze figure of Mushmouse mounted on yellow plastic wagon with pull string. Originally packaged in polybag with illustrated header card. **$20-30**

164

165

164) **MUSHMOUSE AND PUNKIN PUSS SURPRISE PICTURES TO TRACE AND COLOR** (Whitman 1964) 8"x11",40-pages. Each page has an accompanying tracing page. **$15-25**

165) MUSHMOUSE HANDPUPPET (Ideal 1964) 10" puppet with detailed soft vinyl head and cloth body. **$20-35**

168

169

168) PUNKIN' PUSS HANDPUPPET (Ideal 1964) 11" puppet with detailed soft vinyl head and cloth body. **$20-35**

169) PUNKIN' PUSS BUBBLE BATH CONTAINER (Purex 1965) 10" plastic soap container with hard plastic removable head. **$15-25**

170

171

170) PUNKIN' PUSS SQUEEZE FIGURE (Magvi 1967) 6" painted rubber squeeze figure depicts Punkin' Puss standing with shotgun. **$20-35**

171) PETER POTAMUS COLORING BOOK (Whitman 1965) 8"x11",80+ pages. **$15-20**

172

173

172) PETER POTAMUS FRAME TRAY PUZZLE (Whitman 1965) Inlay jigsaw puzzle in 11"x14"frame tray. **$10-20**

173) PETER POTAMUS HALLOWEEN COSTUME (Ben Cooper 1965) 10"x12"box contains mask and one-piece fabric bodysuit with illustration of Peter Potamus on front. **$25-50**

174

173) PETER POTAMUS GAME (Ideal 1964) 10"x20"box. **$75-150**

175

176

175) PETER POTAMUS HAND PUPPET (Ideal 1964) 11" soft vinyl head/cloth body puppet comes in 10"x5"window display box. **$35-45**

176) PETER POTAMUS PLAYING CARDS (Whitman 1965) **$15-20**

177

177) PETER POTAMUS "POP-OUT" TARGET BARREL GAME (Ideal 1964) 5" tall brightly colored plastic can features an extended target of Peter Potamus's portrait which, when hit, releases a soft molded vinyl head of Peter. Plastic spring loaded gun and three darts included. Box is 8"x8"x6". **$75-100**

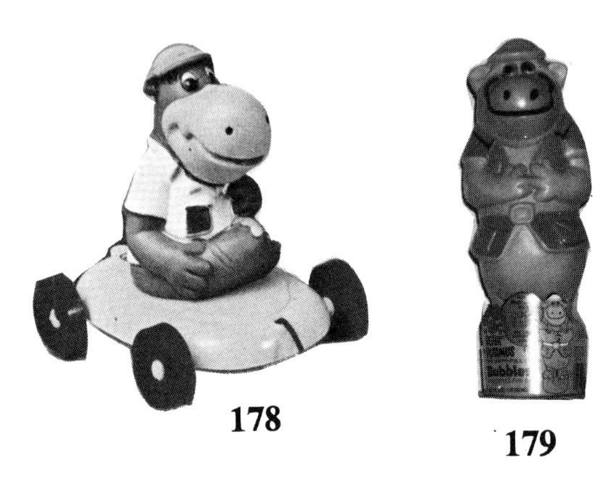

178 179

178) PETER POTAMUS PULL FIGURE (Ideal 1964) 5" painted soft molded vinyl figure of Peter Potamus sitting on plastic wagon with pull string. **$35-50**

179) PETER POTAMUS SOAKY (1965) 10" plastic soap container with hard plastic removable head. **$15-25**

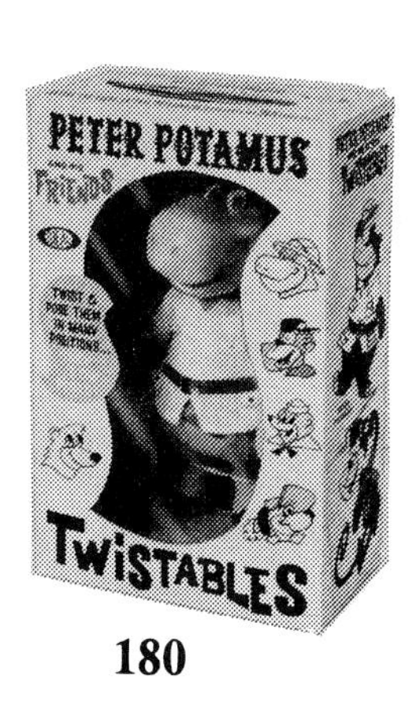

180

181

180) PETER POTAMUS "TWISTABLES" (Ideal 1964) 8" posable stuffed doll with soft vinyl face. Comes in 10"x5" window display box. **$35-50**

181) PIXIE & DIXIE BOOK (Top Top Tales 1961) 6"x8" hardback book with 28-page story accompanied by full color story art on each page. **$12-15**

182

183

182) PIXIE & DIXIE COLORING BOOK (Whitman 1965) 8"x11",28 pages. **$10-12**

183) PIXIE & DIXIE PLUSH DOLLS (Knickerbocker 1960) Pair of two 10" stuffed plush dolls complete with vests, bow-ties and whiskers. **PAIR: $50-75**

184 185 186

184) MR. JINX WITH PIXIE & DIXIE "SOAKY" BUBBLE BATH CONTAINER (Colgate-Palmolive 1963) 10" plastic soap container with hard plastic removable head. Container depicts Mr. Jinx holding Pixie and Dixie. **$10-20**

185) MR. JINX THE CAT CERAMIC FIGURE (1960) Well detailed 5" painted figure. **$35-60**

186) MR. JINX PUSH BUTTON PUPPET (Kohner 1964) 3" painted plastic jointed mechanical figure puppet which moves in a variety of poses when bottom of base is depressed. **$10-20**

187) DIXIE & PIXIE "PUNCH-O" PUNCHING BAG (Kestral 1959) Inflatable 18" tall vinyl plastic bop bag with weighted bottom shows full illustration of Pixie on one side and Dixie on the other. 7"x6"box. **$10-15**

188 **189**

188) MR. JINX PREMIUM FIGURE (Kelloggs 1960) 2.5" hard plastic three-piece figure of Mr. Jinx snaps together at neck and ankles. This figure came free in specially marked boxes of Kelloggs corn flakes. **$20-40**

189) QUICK DRAW McGRAW BOOK (Little Golden Book 1960) 7"x8"hardback book that has full color story art on each page. **$8-12**

190 **191**

190) QUICK DRAW McGRAW AS "EL KABONG" CHILDREN'S RECORD (Golden Records 1960) 6"x8"color stiff paper sleeve contains 78 rpm record with popular El Kabong theme song and "Ooch, Ooch, Ooch." **$25-50**

191) QUICK DRAW McGRAW CHILDREN'S RECORD (Golden Records 1960) 6"x8"color stiff paper sleeve contains 78 rpm record with "Quick Draw's A-Comin' (and Baba Looey, too) to clean up your town" parts A and B. **$25-50**

192

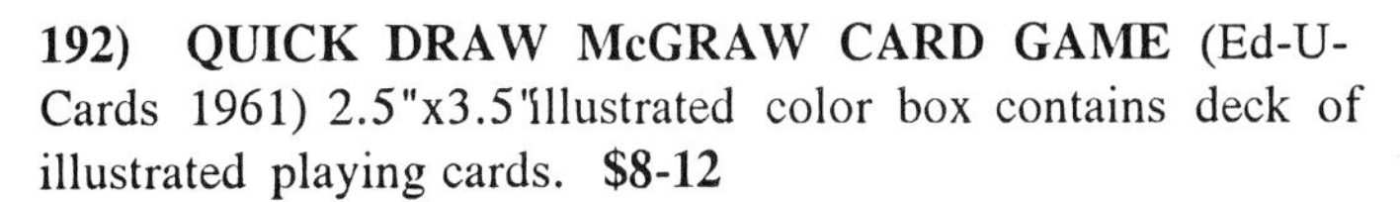

192) QUICK DRAW McGRAW CARD GAME (Ed-U-Cards 1961) 2.5"x3.5"illustrated color box contains deck of illustrated playing cards. **$8-12**

193

193) QUICK DRAW McGRAW COLORING BOOKS (Whitman 1960) EACH: **$8-15**

194 **195**

194) QUICK DRAW McGRAW FIGURAL BANK (Knickerbocker 1960) 10" orange plastic bank with painted accents. **$12-25**

195) QUICK DRAW McGRAW GLOVES (1960) Child's Western-style cloth and fabric gloves with illustration of Quick Draw. **$10-20**

196

196) QUICK DRAW McGRAW MAGIC RUB-OFF PICTURES (Whitman 1960) 8"x16"box contains fourteen 6x8" glossy cardboard slates with black/white illustrations which are colored by the included "Magic Crayons". Wipe-off tissues included so slates may be re-used. **$40-60**

197

197) QUICK DRAW McGRAW "PRIVATE EYE" GAME (Milton Bradley 1960) 10"x20"box contains playing board, small die-cut stand-up character figures of Quick Draw and gang, plus set of playing cards. Object of the game is to solve the mystery. **$25-40**

198) QUICK DRAW McGRAW STUFFED DOLL (Kelloggs 1960) 18" tall stuffed doll originally offered as a premium through boxes of Sugar Smacks cereal. **$20-35**

199

200

199) QUICK DRAW McGRAW AND HUCK HOUND RECORD ALBUM (Golden 1960) 33-1/3 rpm record album contains 16 songs including many theme songs. **$15-30**

200) BABA LOUIE COLORING BOOK (W/S 1960) 8"x11",32-pages. **$8-12**

201) RICOCHET RABBIT CHANGE PURSE (Estelle 1964) 3" diameter white vinyl zipper change purse with embossed color illustration of Ricochet on front. Comes on 6"x4"display card. **$15-25**

202) RICHOCHET RABBIT PUSH BUTTON PUPPET (Kohner 1964) 3" painted plastic jointed mechanical figure puppet which moves in a variety of poses when bottom of base is depressed. **$15-20**

201 202

203

203) RICOCHET RABBIT GAME (Ideal 1965) 10"x20"box. **$100-200**

205

204

204) RICHOCHET RABBIT PULL FIGURE (Ideal 1964) 5" painted soft molded vinyl figure of Ricochet sitting on plastic wagon with pull string. **$35-50**

205) DROOP-A-LONG COYOTE PULL FIGURE (Ideal 1964) 5" painted soft molded vinyl figure of Droop-A-Long sitting on plastic wagon with pull string. **$35-50**

206

206) RUFF & REDDY AT THE CIRCUS GAME (Transogram 1962) 10"x17"box. **$35-45**

207

207) RUFF & REDDY DRAW & COLOR CARTOON SET (Wonder Art 1959) 14"x9"colorful boxed set of comic strip drawings and carbons that allow the user to "trace" a drawing. **$35-65**

208

208) RUFF & REDDY KARBON KOPY CARTOON SET (Wonder Art 1960) 18"x12"box contains tracing set which consists of several color comic strip drawings resembling comic book pages and several sheets of carbon paper. Also included is a 8"x10"sheet of 16 sequence illustrations that can be cut out and placed in order to make a flip-the-page moving picture book. **$75-125**

209

209) RUFF & REDDY MAGIC ERASABLE PICTURES (TransOgram 1958) Nice 15"x10"boxed set, one of the first items made on Ruff and Reddy. **$98**

210 **211**

210) RUFF & REDDY "REDDY" STUFF DOLL (Knickerbacher 1960) 15" stuffed plush doll with hard plastic face. Comes with red collar with "Reddy" embossed in white letters. Illustrated 3"x4"tag attached to collar features Ruff & Reddy on scooter above their name logo. **$75-100**

211) RUFF & REDDY TRAY PUZZLE (Whitman 1959) Inlay jigsaw puzzle in a 11"x14" frame tray with a roller coaster scene. **$12-20**

212

213

212) RUFF & REDDY TV FAVORITES SPELLING GAME (Milton Bradley 1959) 6"x9"box. **$20-30**

213) **RUFF & REDDY CHILD'S RECORD** (1959) 45 rpm record in illustrated paper sleeve. **$10-15**

214

214) **SCOOBY DOO STUFFED DOLL** (J.S. Sutton and Sons 1970) 14" tall stuffed replica of Scooby Doo, made in light orange/brown with dark brown accents. 3.5" illustrated tag of Scooby reads "Scooby Doo where are you!" **$25-50**

215

217

215) **SCOOBY DOO TALKING VIEW-MASTER REELS** (GAF 1970) 8"x8" boxed set contains reels and accessories. **$8-12**

216) **SECRET SQUIRREL TRAY PUZZLE** (Whitman 1965) Inlay jigsaw puzzle in a 11"x14" frame tray. **$10-20**

217) **SECRET SQUIRREL TRICKY TRAPEZE** (Kohner 1964) 3" plastic jointed mechanical figure is suspended from trapeze bar which is mounted to a base with a button on each side. When buttons are squeezed, the figure flips over the trapeze. **$20-25**

218

218) **SNAGGLEPUSS BOARD GAME** (Transogram 1962) 10"x17" box. **$35-50**

219

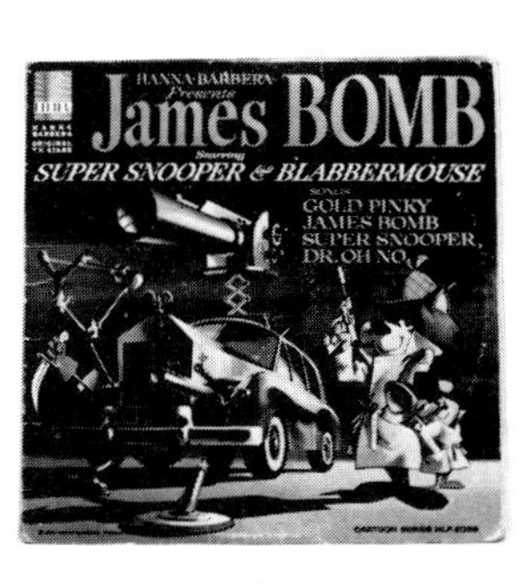

220

219) **SNOOPER AND BLABBER AND QUICK DRAW McGRAW BOOK** (Whitman 1960) 5"x7" hardback book contains 80+ pages of comic book-style stories. Part of the Whitman Comic Book series. **$12-18**

220) **JAMES BOMB STARRING SUPER SNOOPER & BLABBERMOUSE** (HBR 1965) 33-1/3 rpm record contains songs and stories spoofing the spy films. **$15-25**

221

222

221) **SPACE GHOST COLORING BOOK** (Whitman 1965) 8"x11", 80+ pages. **$20-30**

222) **SPACE GHOST JIGSAW PUZZLE** (Whitman 1967) 8"x10" box contains 100 piece puzzle. **$20-25**

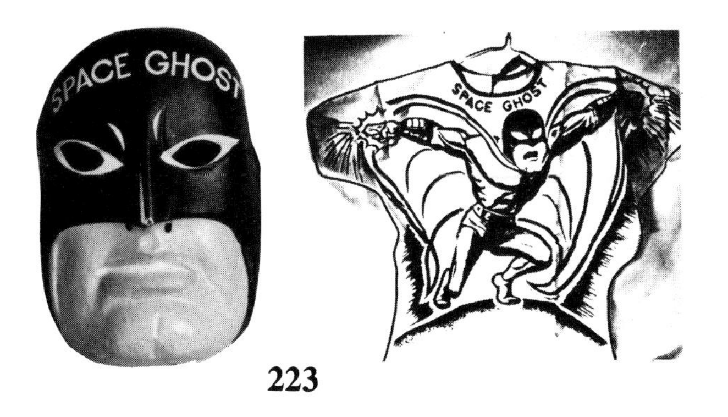

223

223) SPACE GHOST HALLOWEEN COSTUME (Ben Cooper 1965) 10"x12"box contains mask and one-piece fabric bodysuit with illustration of Space Ghost on front. **$50-100**

224

225

224) SPACE GHOST & FRANKENSTEIN JR. BUBBLE BATH (Purex 1966) 9"x6" color illustrated box features Hanna-Barbera tee-shirts and nightlights on back of box. **$50-100**

225) SPACE KIDETTES MAGIC SLATE (Watkins/Strathmore 1967) 8"x12" illustrated cardboard display card holds lift-up erasable film sheet and comes with wood stylus. **$15-30**

226) SPACE KIDETTES JIGSAW PUZZLE (Whitman 1967) 8"x10" box contains 70-piece jigsaw puzzle which assembles into a 14"x18"scene of the Space Kidettes. **$15-30**

227

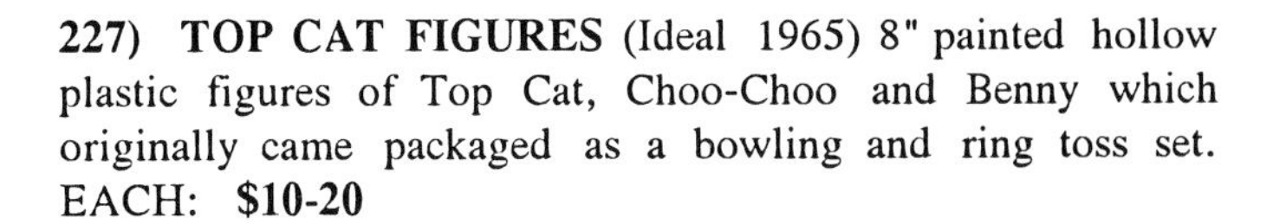

227) TOP CAT FIGURES (Ideal 1965) 8" painted hollow plastic figures of Top Cat, Choo-Choo and Benny which originally came packaged as a bowling and ring toss set. EACH: **$10-20**

228 229

228) TOP CAT COLORING BOOK (Whitman 1961) 8"x11", 100-pages. **$15-25**

229) TOP CAT GAME (Whitman 1962) Game contains colorful 14"x14"rotating notched board which shows an aerial view of a city. Game contains discs depicting milk, newspaer and fish, two spinners and several marble playing pieces. Object of the game is to be the first player to acquire all three disc subjects and arrive at Top Cat's trash can home. **$50-75**

230

230) TOP CAT IN ROBIN HOOD RECORD (HBR 1962) Colorful 6"x6"stiff paper sleeve contains 45 RPM record with four songs including "Top Cat" song. **$10-20** 1977 LP: **$5-7**

231

231) TOP CAT JIGSAW PUZZLE (Whitman 1961) 10"x12" box contains 70-piece jigsaw puzzle which assembles into colorful scene of Top Cat and gang playing baseball. **$20-30**

232 233

232) TOP CAT SOAKY (1963) 10" plastic soap container with hard plastic removable head. **$15-30**

233) TOP CAT TRAY PUZZLE (Whitman 1961) Inlay jigsaw puzzle in a 11"x14"frame tray. **$12-15**

234 235

234) TOUCHE TURTLE DRIVE TOY (Marx 1965) Small yellow/green hard plastic figure of Touche Turtle with white hat and accented facial features. Figure has hidden wheels which are friction drive and feet which move back and forth when the toy is in motion. Box is 4"x4"x4".**$50-75**

235) DUM DUM & TOUCHE TURTLE DRIVE TOY (Marx 1965) 3"hard plastic figure of Dum Dum riding on top of Touche Turtle, who is designed more like a scooter than a turtle. Friction drive action allows toy to roll across surfaces. Box is 4"x4".**$35-50**

236

236) TOUCHE TURTLE, LIPPY THE LION & WALLY GATOR CUP (Gothem Ind. 1964) Plastic drinking cup shows all three characters marching with band instruments. Illustration is on paper sealed inside clear plastic coating of cup. **$15-25**

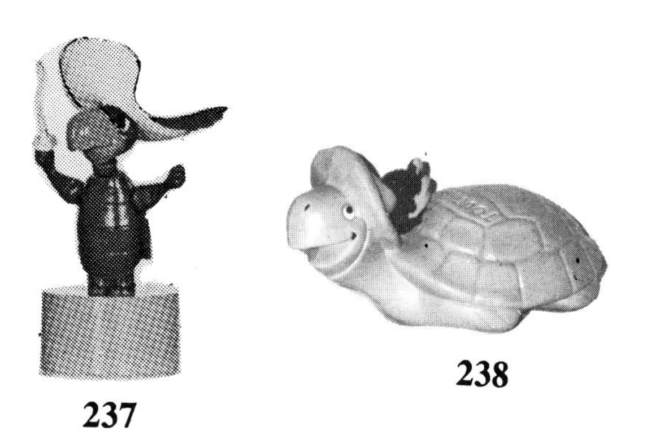
237 238

237) TOUCHE TURTLE PUSH BUTTON PUPPET (Kohner 1964) 3" painted plastic jointed mechanical figure puppet which moves in a variety of poses when bottom of base is depressed. **$15-25**

238) TOUCHE TURTLE SOAKY (1963) 10" plastic soap container with hard plastic removable head. **$20-30**

239) TOUCHE TURTLE GAME (Ideal 1964) 10"x20"box. **$100-150**

240

240) WACKY RACES BOARD GAME (MB 1968) $25-50

241

241) **WACKY RACES CARS** (Hasbro 1969) 3"long detailed plastic replica cars driven by Dick Dastardly's Mean Machine, Penelope Pitstop's Compact Pussycat and Peter Perfect's Turbo Terrific, with detailed figure of each character behind the wheel. Each car comes individually packaged on a 4"x6" oval-shaped display card. EACH: **$25-40**

242

243

242) **WACKY RACES FRAME TRAY PUZZLE** (Whitman 1969) 11"x14"tray puzzle of Dick Dastardly and Muttley about to crash into two other bi-planes. **$10-15**

243) **WAÇKY RACES JIGSAW PUZZLE** (Whitman 1970) 10"x12"box contains 70-piece puzzle which assembles into a 14"x18"racing scene of Luke and Blubber Bear and Sawtooth. **$15-20**

244) **DICK DASTARDLY & MUTTLEY GAME** (Milton Bradley 1969) 9"x16"box. **$25-40**

245) **DICK DASTARDLY POWERFUL MAGNETS** (Larami 1970) 4"x7"illustrated display card contains a large and small plastic horseshoe-shaped magnet. **$10-15**

245

246

246) **PENELOPE PITSTOP JEWELRY** (1969) 5"x6"display card contains an assortment of plastic facsimile gems. **$8-10**

247

249

247) **WACKY RACES "COMPACT PUSSYCAT" MODEL CAR KIT** (MPC 1969) 9"x5"box contains all-plastic assembly kit of Penelope Pitstop's "Compact Pussycat" race car. Molded in metallic purple and yellow. **$75-125**

248) **WALLY GATOR GAME** (Transogram 1962) 10"x17" box. **$45-75**

249) **WALLY GATOR PUSH BUTTON PUPPET** (Kohner 1964) 3" painted plastic jointed mechanical figure puppet which moves in a variety of poses when bottom of base is depressed. **$15-25**

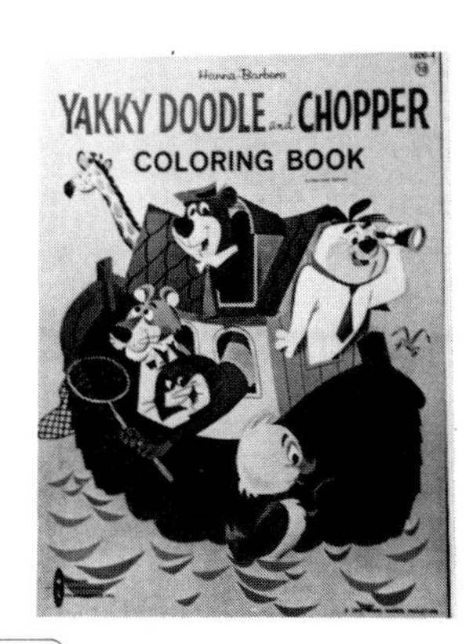

250

251

250) YACKY DOODLE AND CHOPPER COLORING BOOK (Watkins-Strathmore 1962) 8"x11",30-pages. **$10-12**

251) YOGI BEAR "BREAK-A-PLATE" GAME (Transogram 1960) Large 18"x13"boxed set contains three illustrated moving break-apart plates encased in a target range setting, and plastic balls to throw at plates. **$25-50**

252

252) YOGI BEAR CARTOONIST STAMP SET (Lido 1961) Large 18"x10"window display box holds several stamps of H-B characters, stamp pad, etc. **$20-35**

253

254

253) YOGI BEAR COOKIE JAR (1960) Colorful ceramic 15" figure of Yogi, complete with felt tongue that is usually missing. Sign beside him reads: "Better that the Average Cookie". **$150-250**

254) YOGI BEAR CRAYON BY NUMBER AND STENCIL SET (TransOgram 1963) 9"x16"window display box contains cut-out stencils of Yogi and gang, eight pre-numbered sketches to color, 24 crayons and sharpener. **$35-50**

255

256

255) YOGI BEAR "BUBBLE" PIPE (Transogram 1963) 5"x7"display card holds plastic pipe with Yogi's head forming the front of the pipe. **$10-15**

256) YOGI BEAR FIGURAL BANK (Knickerbocker 1959) Plastic 10" bank of Yogi comes in package with illustrated header card. **$15-25**

257

257) YOGI BEAR "GO FLY A KITE" GAME (Transogram 1961) 10"x20"box. **$25-45**

258

258) YOGI BEAR HEAD MASK (Kelloggs 1961) Large fold-together color cardboard pull over head mask available as a mail order premium through Kelloggs Corn Flakes. **$35-50**

259

259) **YOGI BEAR, HUCKLEBERRY HOUND AND SNAGGLEPUSS MAGIC TRACER** (Whitman 1962) 20"x10" box contains 14 colorful 8"x10"illustrations of Huck, Yogi and Snagglepuss. Illustrations are placed between two cardboard holders and large plastic reflector and traced and then colored. Set includes a box of crayons and drawing paper, plus three sheets which show how to draw each character in easy to follow steps. **$50-100**

260

260) **YOGI BEAR JELLYSTONE NATIONAL PARK PLAYSET** (Marx 1962) Attractive 15"x22" illustrated box contains eight 2.5" painted figures of Yogi, Boo-Boo, Cindy Bear, Ranger Smith, Snagglepuss, and other Hanna-Barbera characters. Set also includes ranger station, jeep, terrain, trees, gate entrance, park animals and several other accessories. **$400-600**

261

261) **YOGI BEAR MAGIC RUB-OFF PICTURES** (Whitman 1961) 8"x16"box contains 14 glossy 6"x8"cardboard slates with black/white illustrations of Hanna-Barbera characters which are colored, wiped clean and re-colored with included crayons and tissues. **$50-60**

262 263

262) **YOGI BEAR MAGIC SLATE** (1960) 8"x12"illustrated cardboard display card holds lift-up erasable film sheet and comes with wood stylus. **$15-25**

263) **YOGI BEAR STUFFED DOLL** (Knickerbocker 1959) 10" stuffed doll with vinyl face. **$25-50**

264 265

264) **CINDY BEAR PUSH BUTTON PUPPET** (Kohner 1964) 3" painted plastic jointed mechanical figure puppet which moves in a variety of poses when bottom of base is depressed. **$10-20**

265) **YOGI BEAR PUSH BUTTON PUPPET** (Kohner 1964) 3" painted plastic jointed mechanical figure puppet which moves in a variety of poses when bottom of base is depressed. **$10-20**

HARVEY CHARACTERS

Many of the Harvey characters known to us today were originally created by famous studios for Paramount Pictures and shown as theatrical cartoons before the main feature film. Animation producer Joseph Oriolo created Casper the Friendly Ghost for Paramount in 1946 for the sum of $175. Little Audrey was created in 1947 to replace Little Lulu, which was created by Fleischer Studios and for which Paramount had lost the rights. Baby Huey was born in 1951. These cartoons were all in syndication to TV networks in the Fifties and Sixties. 26 new cartoon episodes of Casper and his supporting cast were produced in 1963 for the half hour program "The New Casper Show" which ran on ABC throughout the Sixties.

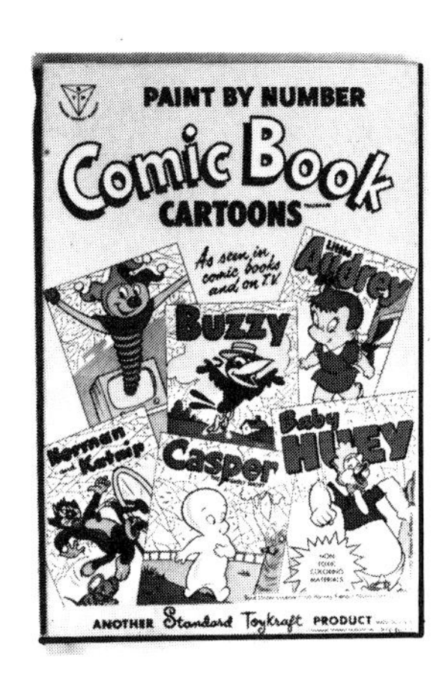

1

2

1) HARVEY "COMIC BOOK CARTOONS" SETS (Standard Toykraft 1960) 10"x14"fold-open box is designed like a book and contains twelve pre-numbered sheets featuring Harvey cartoon characters and a free Harvey comic book. Three different sets were produced, which included either paints, crayons or color pencils. EACH: **$35-75**

2) HARVEYTOON TINKLE TOY "STUFFIES" (Saalfield 1960) Boxed set of six stuffed Harvey characters averaging 5" to 6" in height. Each character comes with a front and back and a stuffed middle and is laced together by colored yarn including bells. Characters include Casper, Wendy, Spooky, Baby Huey, Herman and Katnip and Little Audrey. **$75-100**

3

3) **HARVEYTOON TV FLANNEL FUNNIES** (Saalfield 1960) Colorful 11"x14"boxed set contains large TV-shaped board on which Harvey flannel figures and accessories are placed to create a scene (like Colorforms). Characters include Casper, Wendy, Spooky, Baby Huey, Little Audrey, Herman and Katnip. **$50-60**

4

4) **FUN DAY CARTOONS CARD GAME** (Built-Rite 1959) Small deck of illustrated die-cut shaped cards come in 3"x4" illustrated cardboard box and is played like the game Authors. **$8-12**

5

6

5) **BABY HUEY AND PAPA BOP BAG** (Doughboy 1966) 54" tall inflatable vinyl plastic bounce-back bop bag with weighted, rolled bottom. **$50-65**

6) **BABY HUEY COLORING BOOK** (Saalfield 1959) 8"x11",80-pages. **$10-15**

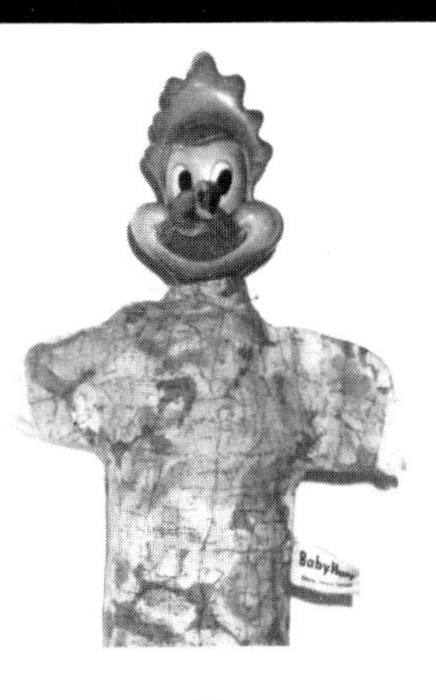

7

8

7) **BABY HUEY HAND PUPPET** (Gund 1960) 11" handpuppet with molded vinyl head and fabric body. **$20-25**

8) **BABY HUEY SQUEEZE FIGURE** (1960) 10" soft vinyl squeeze figure with color accents depicting Baby Huey holding an ice cream cone. **$40-75**

9

10

9) **BABY HUEY JIGSAW PUZZLE** (Built-Rite 1961) 8"x9" box contains 70-piece puzzle. **$15-20**

10) **BABY HUEY BOOK** (Wonder Book 1960) 7"x8" hardback book with 20+ page story accompanied by full color story art on each page. **$10-20**

11

20

20) CASPER THE TALKING GHOST DOLL (Mattel 1961) 15" stuffed doll with hard plastic face has pull-string in back of neck which allows doll to produce a variety of phrases. Doll comes in large 8x16x6"purple box designed like a haunted house with front and side windows. **$125-175**

21

22

21) CASPER THE FRIENDLY GHOST CHIDREN'S RECORD (Little Golden 1959) 78 rpm record comes in color paper sleeve and also features Little Audrey. **$8-12**

22) CASPER THE FRIENDLY GHOST GE-TAR (Mattel 1960) 20" tall, black hard plastic guitar has full color die-cut lithographed paper label depicting Casper reading a song book. There is a built-in music box and when the crank on the side of the guitar is turned, a melody is produced. Box is 20"x6"x2".**$25-50**

23) CASPER THE GHOST ED-U-CARD (1961) 2.5"x3.5" card set containing many of the Harvey gang. **$8-12**

24

24) CASPER THE GHOST BOARD GAME (MB 1959) Object of the game is to help Casper find his way home. **$10-15**

25) CASPER HALLOWEEN COSTUME (Ben Cooper 1961) 10"x12"box contains mask and one-piece fabric bodysuit with illustration of Casper on front. **$20-25**

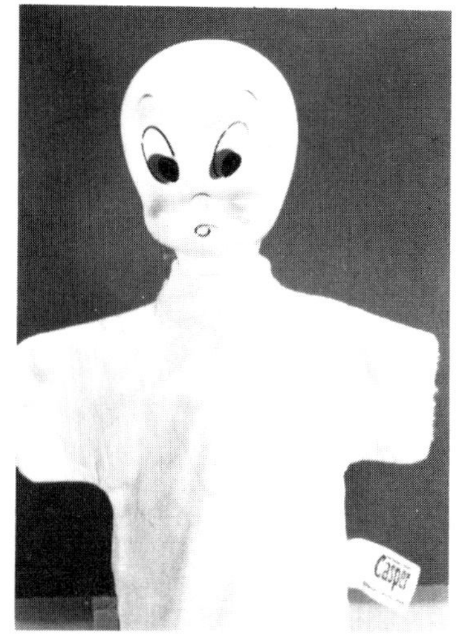

26

27

26) CASPER HAND PUPPET (Gund 1960) 10" puppet with cloth body and soft vinyl head. **$20-25**

27) CASPER HAUNTED HOUSE BALLOON (Van Dam 1960) 10"x6"display card contains scary house and Casper balloon. **$20-25**

28) CASPER "HI-C" DRINK ADVERTISING POSTER (1970) Large 21"x13"colorful poster of Casper promoting the popular canned drink. **$25-50**

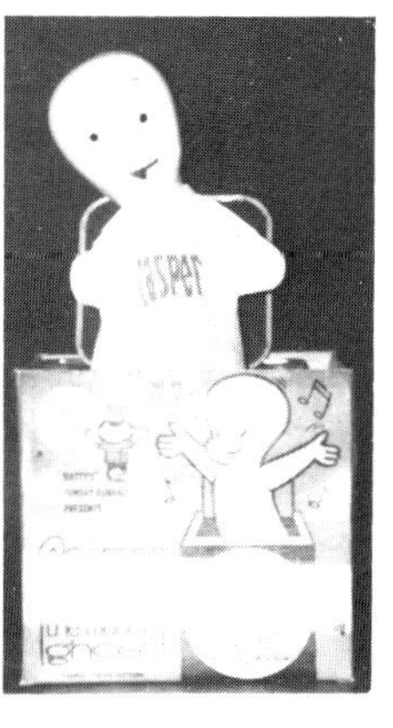

29

30

29) CASPER JACK IN THE BOX (Mattel 1961) 5"x5"x6" color litho metal crank-wound music box contains pop-up Casper figure with plastic head and cloth body. **$50-75**

11) **BUZZY THE CROW JIGSAW PUZZLE** (Built-Rite 1961) 8"x9"box contains 70-piece puzzle. **$15-25**

12

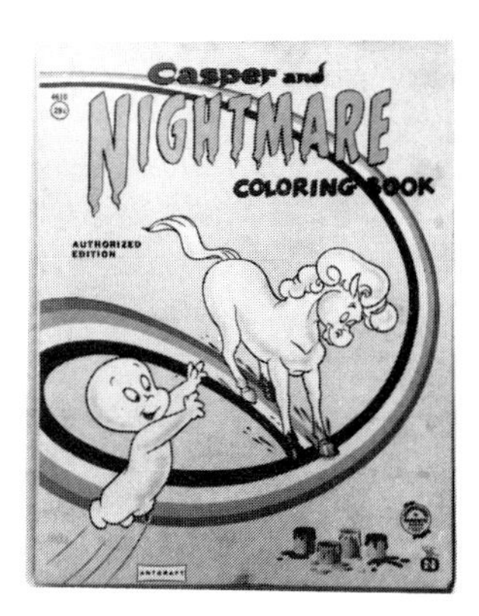

13

12) **BUZZY THE FUNNY CROW BOOK** (Wonder Books 1960) 7"x8"hardback book with 20+ page story with full color story art on each page. **$8-10**

13) **CASPER AND NIGHTMARE COLORING BOOK** (Saalfield 1964) 8"x11",100-pages. **$15-25**

14

15

14) **CASPER BALLOON FIGURE** (1962) 9"x8"display card holds large Casper-shaped balloon. **$20-25**

15) **CASPER CERAMIC BANK** (late 1940's) Nicely made 8" ceramic figure of Casper holding a large bag of money. **$75-100**

16) **CASPER THE GHOST MUSICAL DOLL** (Knickerbocker 1960) 11"stuffed musical wind-up doll dressed in blue vest and cap that plays a baby's lullaby. **$25-30**

16

17

17) **CASPER ELECTRONIC ADVENTURE GAME** (Tarco 1962) 15"x11"x3"deep box contains built-in playing board depicting a rainbow path running through dangerous pitfalls like a dragon's den, ice caves, dismal swamp, etc. Game has hidden built-in "remote control" magnetic device with steering stick to guide magnetic plastic figure of Casper across the board. Also included are four plastic mountain tunnels and pot of gold. Object of the game is to travel safely through the hazardous terrain and be the first to bring the pot of gold back. **$100-200**

18

18) **CASPER THE GHOST CEREAL BOWL AND MUG SET** (Westfield 1960) 5" diameter x 2" deep milk white glass cereal bowl features wrap around illustrations of Casper, Wendy, Spooky and Nightmare, the horse, against a bright orange background. The mug features the same illustrations against an orange background. SET: **$25-40**

19

19) **CASPER THE GHOST LUNCHBOX** (King Seeley Thermos 1966) Blue vinyl box with orange steel thermos. Box: **$200-300** Thermos: **$25-75**

30) CASPER "JUMPING BEAN" GAME (Milton Bradley 1959) 6"x10"window display box contains 3-in-one game board and magic jumping beans. **$15-25**

31

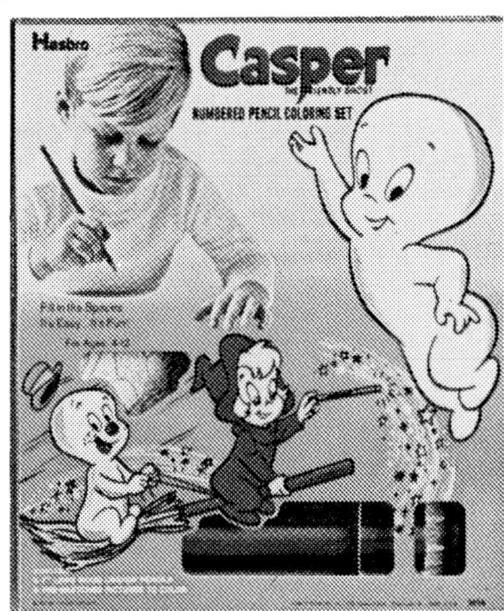

32

31) CASPER LAMPSHADE (1960) Colorful 10" diameter lampshade featuring all the Casper characters. **$25-45**

32) CASPER NUMBERED PENCIL COLORING SET (Hasbro 1961) 8"x10" box contains eight pre-numbered sketches, twelve color pencils and sharpener. **$25-50**

33

33) CASPER RUB-A-PENCIL SET (Saalfield 1960) 9"x12" box contains five coloring books of the same size which feature pages to color plus individual black "magic pages" which when rubbed with a pencil will produce a surprise picture. Set includes pencil. **$35-50**

34

35

34) CASPER PAINT BY NUMBER 'N FRAME SET (Hasbro 1961) 12"x10"window display box contains four pre-numbered sketches, twelve watercolor paints, brush and blue plastic frame. **$25-50**

35) CASPER PARTY SET (Reed 1965) 6 piece set contains tablecloth, cups, napkins, small and large plates. **$20-30**

36

36) CASPER PICTURE LOTTO (Built-Rite 1959) 18"x9" box contains several cardboard lotto sheets, each illustrated with nine Harveytoon characters, and matching character tiles. **$15-25**

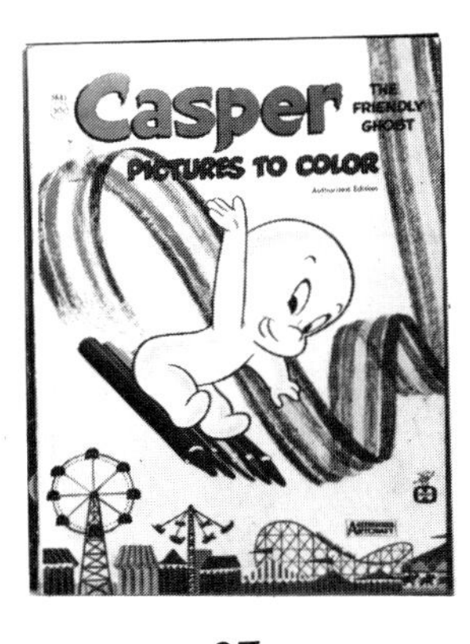

37

38

37) CASPER "PICTURES TO COLOR" COLORING BOOK (Saalfield/Artcraft 1960) 8"x11"deluxe, 300+ page coloring book. **$15-35**

38) CASPER SLIDE TILE PUZZLE (Roalex 1961) 8"x5" display card holds 3"x3"b/w plastic puzzle with movable tiles to form image of Casper. **$25-35**

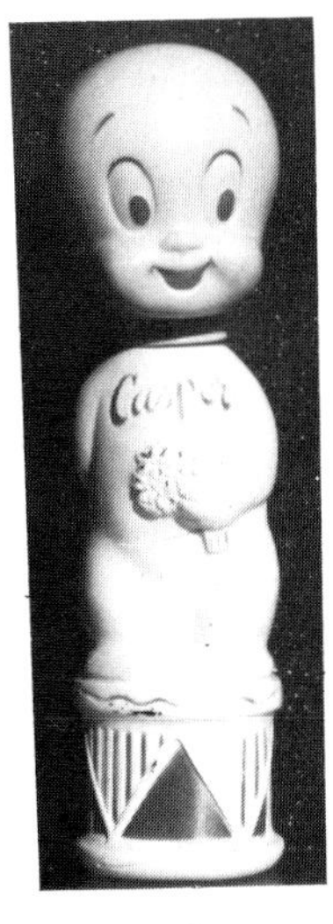

39

40

39) CASPER SOAKY BUBBLE BATH CONTAINER (1962) 10"plastic soap container with hard plastic removable head. **$15-20**

40) CASPER TIN WIND-UP TOY (Line Mar 1959) 5" tall tin litho figure of Casper which, when wound, hops about as his head bounces up and down. **$150-250**

41) CASPER THE GHOST VIEWMASTER REEL SET (1961) 5"x5"color illustrated envelope contains three reels and 16-page story booklet. **$10-12**

41

42

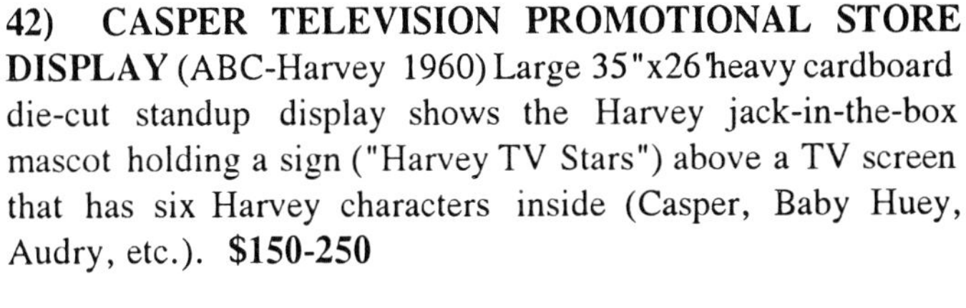

42) CASPER TELEVISION PROMOTIONAL STORE DISPLAY (ABC-Harvey 1960) Large 35"x26"heavy cardboard die-cut standup display shows the Harvey jack-in-the-box mascot holding a sign ("Harvey TV Stars") above a TV screen that has six Harvey characters inside (Casper, Baby Huey, Audry, etc.). **$150-250**

43

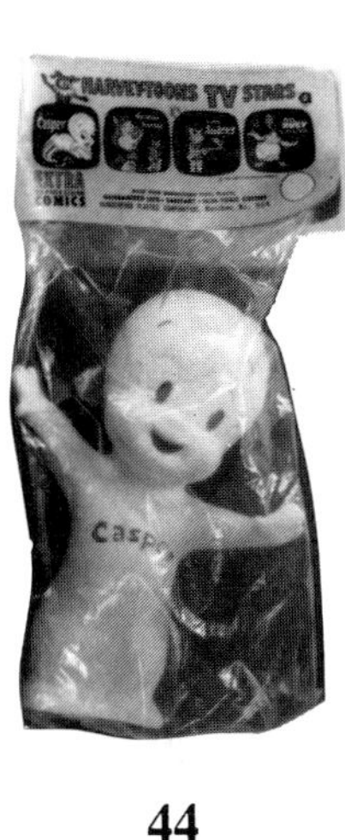

44

43) CASPER THE GHOST VINYL SQUEEZE FIGURE (1960) 8" soft vinyl squeeze figure depicts Casper holding a Dalmation puppy. **$50-100**

44) CASPER VINYL FIGURE (Hungerford 1960) 8" white vinyl figure of Casper with painted accents on face and his name across his chest. Figure comes in poly bag and the header card is actually a small 20-page comic book. **$25-50**

45

45) CASPER'S "TREAT TOTE" HALLOWEEN BAG (1970) 18"x15" thin plastic carry bag featuring Casper holding the same bag (infinity-style design) in colors of orange, black and white. **$15-30**

46

47

46) CASPER AND WENDY BOP BAG (Doughboy 1966) 54" tall inflatable vinyl plastic bounce-back bop bag with weighted, rolled bottom. **$40-60**

47) CASPER & WENDY KITE (Saalfield 1960) Punch-out cardboard kite comes on attractive 9"x11" colorfully illustrated format. **$10-12**

48

49

48) THE GHOSTLY TRIO COLORING BOOK (Saalfield/Artcraft 1961) 8"x11" 100-page. **$15-25**

49) HERMAN & KATNIP COLORING BOOK (Saalfield/Artcraft 1960) 8"x11", 80+ pages. **$15-25**

50) KATNIP TIN WIND-UP TOY (Line Mar 1959) 5" tall tin litho figure of Katnip which, when wound, hops about as his head bounces up and down. **$200-300**

50 **51**

51) HERMAN TIN WIND-UP TOY (Line Mar 1959) 5" tall tin litho figure of Herman which, when wound, hops about as his head bounces up and down. **$200-300**

52

53

52) HERMAN AND KATNIP BOP BAG (Doughboy 1966) 54" tall inflatable vinyl plastic bounce-back bop bag with weighted, rolled bottom. **$40-60**

53) HERMAN & KATNIP DEEP VIEW PAINT SET (Pressman 1961) 11"x14"x2" box contains 11"x14" framed 3-D picture with painted background. Comes with six paints, water bowl and brush. **$50-75**

54

54) HERMAN & KATNIP GAME BOX (Saalfield/Artcraft 1960) 18"x9"box contains playing board with a game on each side, 24 die-cut playing pieces and dice. **$25-45**

55) KATNIP & HERMAN KITE (Saalfield 1960) Interesting punch-out cardboard kite comes on attractive 9"x11"colorfully illustrated format. **$15-25**

56

57

56) HOT STUFF CERAMIC FIGURE (1960's) 7" painted glazed ceramic figure depicts Hot Stuff holding a plastic pitchfork. Name on base and copyright on back of base. **$50-100**

57) LITTLE AUDREY CHILDRENS RECORD (1960) 45 rpm record in paper sleeve. **$10-15**

58

58) LITTLE AUDREY "DOES THE DISHES" KITCHEN SET (1960) Large 35"x15"x12"colorful boxed set contains miniature kitchen sink, dish rack, and food products. **$50-75**

59) LITTLE AUDREY HALLOWEEN COSTUME (Collegeville 1959) 10"x12"box contains mask and one-piece fabric bodysuit with illustration of Little Audrey on front. **$25-50**

59

60

60) LITTLE AUDREY HAND PUPPET (Gund 1960) 12" handpuppet with molded vinyl head and fabric body. **$25-35**

61

62

61) LITTLE AUDREY TIN WIND-UP TOY (Line Mar 1959) 5" tall tin litho figure of Little Audrey which, when wound, hops about as his head bounces up and down. **$150-250**

62) LITTLE AUDREY SHOULDER BAG LEATHER CRAFT KIT (Jewel Leathergoods Co. 1961) 16"x11"box contains six leather items which are to be sewn together and made into two purses, comb case, bell key caddy, picture frame and other items. Set includes plastic lacing and lacing instructions. Features illustrations of Little Audrey inside the box. **$50-75**

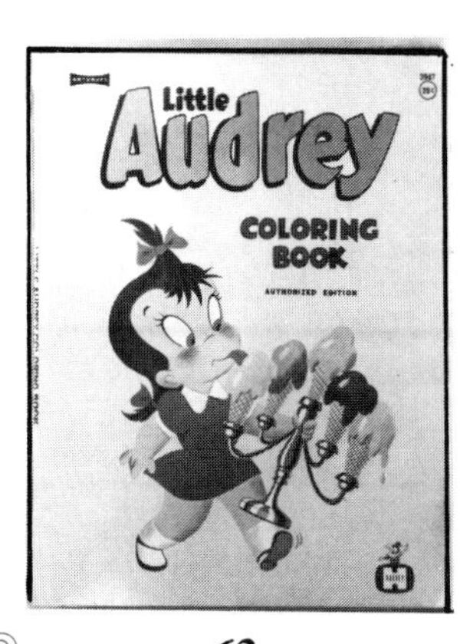

63

64

63) LITTLE AUDREY TV COLORING BOOK (Saalfield 1962) 8"x11",80-pages. **$15-25**

64) LITTLE AUDREY COLORING BOOK (Saalfield 1960) 8"x11",80+ pages. **$15-20**

65

65) LITTLE AUDREY VINYL CARRY BAG (1960) "Around the World with Little Audrey" **$25-50**

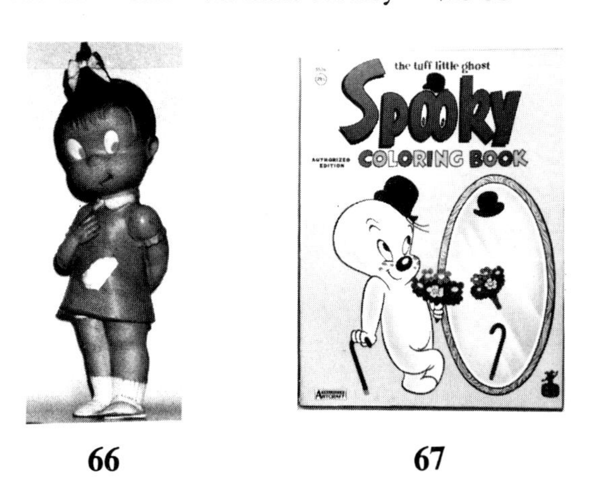

66 67

66) LITTLE AUDREY VINYL FIGURE (1950's) Large 12" colorfully detailed figure depicts Audrey in blue dress. **$50-75**

67) SPOOKY THE TUFF LITTLE GHOST COLORING BOOK (Saalfield/Artcraft 1959) 8"x11",80+ pages. **$15-30**

68

68) SPOOKY THE TUFF LITTLE GHOST KARBON KOPEE SET (Wonder Art 1960) 8"x12"box contains seven color carbon copy papers to "draw your own TV funnies," drawing board, pencil and wood stylus. Also included is an eight-page color booklet showing various covers of spooky comic books published by Harvey comics, plus a 32-page book which contains the same comic cover illustrations in b/w and blank pages to trace in. There is also included a 6"x8"sheet of twelve sequence illustrations that can be cut out in order to make a flip-page moving picture booklet. **$50-100**

69

69) SPOOKY AND CASPER DEVILS FOOD SANDWICH COOKIES (1968) 10"x5"x3"box with pop-up handle contains Oreo-style chocolate cookies with cream centers. Each cookie has an embossed image of Spooky or Casper's head. Limited distribution. **$100-200**

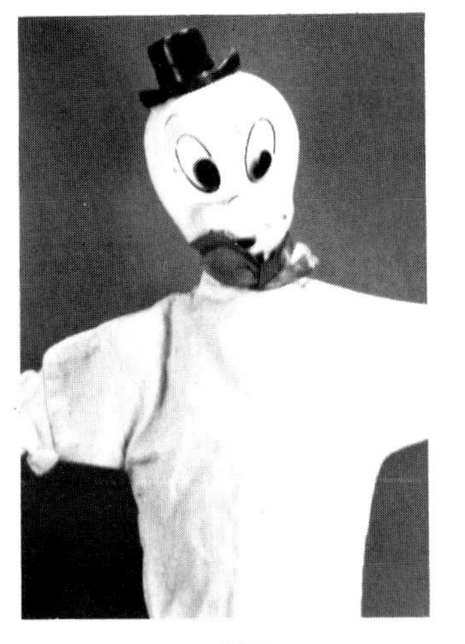

70

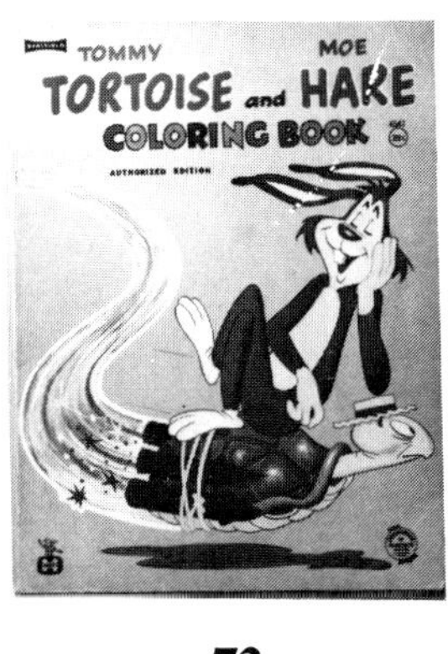

72

70) SPOOKY HAND PUPPET (Gund 1960) 11"handpuppet with molded vinyl head and fabric body. **$25-35**

71) SPOOKY STUFFED DOLL (Gund 1960) 8" stuffed doll with hard vinyl head. **$35-50**

72) TOMMY TORTOISE & MOE HARE COLORING BOOK (Saalfield 1959) 8"x11",80+ pages. **$15-25**

73

74

73) TOMMY TORTOISE & MOE HARE JIGSAW PUZZLE (Built-Rite 1961) 8"x9"box contains 70-piece puzzle. **$15-25**

74) WENDY AND HER MAGIC WAND STORY BOOK (McGraw-Hill 1964) 9"x11"thick cardboard book with die-cut cover and spiral binding contains 16-page story accompanied by full color story art. **$15-25**

75

75) WENDY BOARD GAME (MB 1966) **$50-65**

76

77

76) WENDY HAND PUPPET (Gund 1960) 12"handpuppet with molded vinyl head and fabric body. **$25-50**

77) WENDY HALLOWEEN CANDY BUCKET (1973) Large plastic bucket in the shape of Wendy's head. **$10-12**

78

78) WENDY TRAY PUZZLE (Built-Rite 1960) Inlay jigsaw puzzle in a 11"x14"frame tray. **$15-20**

79) WENDY SOAKY BUBBLE BATH CONTAINER (1963) 10" plastic soap container with hard plastic removable head. **$20-25**

80

80) WENDY THE GOOD LITTLE WITCH COLORING BOOKS (Saalfield/Artcraft 1959-62) 8"x11",80-100 pages. EACH: **$10-20**

JAY WARD CHARACTERS

Premiering on ABC TV on September 29, 1959, the "Rocky and his Friends" cartoon show ran only two years. NBC picked up the cartoon and re-ran it under "The Bullwinkle Show" for one year (1961-62), and after only moderate success, tossed it back to ABC which began re-running it again in 1964 through 1971.

What started as a shaky beginning for Jay Ward and his creations Bullwinkle and Rocky has grown into almost legendary pop culture, and has retained its popularity over the years. The Cold War overtones, cloaked adult humor, and witty puns have made Jay Ward cartoons a treasured piece of the baby boom era. Supporting characters Dudley Do-Right, Mr. Peabody and his pet boy Sherman, and Boris and Natasha have all gained cult followings.

Ward's other cartoon series credits include "George of the Jungle" (1967-70), which also featured "Super Chicken" and "Tom Slick". Ward also produced the Cap'n Crunch, Quisp and Quake commercials for the Quaker Oats Company in the 1960's and also co-produced many cartoons with smaller animation companies.

1

2

3

4

1) BULLWINKLE AND ROCKY BOARD GAME (Ideal 1963) 10"x20"box contains playing board, pieces, etc. Object of the game is to be the first to pass through hidden obstacles and find hidden treasure. **$25-50**

2) BULLWINKLE AND ROCKY COLORING BOX (Whitman 1960) 8"x12"box contains 256 coloring book pages and crayons. **$25-40**

3) BULLWINKLE CARTOON KIT (Colorforms 1962) 8"x12"box contains thin vinylplastic character pieces that stick to illustrated background board. **$75-125**

4) BULLWINKLE DRINKING GLASS (Pepsi 1973) Color 5" tall glass of Bullwinkle running after a balloon at an amusement park. **$12-15**

5 6

5) **BULLWINKLE COLORING BOOK** (Whitman 1965) 8"x11",80-pages. **$15-30**

6) **BULLWINKLE ELECTRIC QUIZ GAME** (Larami 1971) 7"x12"illustratd display card contains colorful board with 24 different scenes, contact maker with lighting bulb, plus extra quiz boards. **$15-20**

7 8

7) **BULLWINKLE JIGSAW PUZZLE** (Whitman 1972) 10"x12"box contains 100-piece jigsaw puzzle which assembled into a 14"x18"scene depicting Rocky and Bullwinkle catching Boris and Natasha burglarizing a safe. **$10-20**

8) **BULLWINKLE FIGURAL RADIO** (1969) 12"tall plastic figural radio depicts Bullwinkle in a red/white full piece swim suit and white gloves with yellow antlers and feet. Control dials are on his back. **$100-150**

9) **BULLWINKLE FOR PRESIDENT BUMPER STICKER** (1972) 12"x4"sticker reads, "He's Alot of Bull". **$10-15**

10 11

10) **BULLWINKLE HIDE 'N' SEEK GAME** (Milton Bradley 1961) 10"x19"box contains two illustrated panel playing boards and 48 playing discs. Object is to find the 16 discs with characters. **$50-75**

11) **BULLWINKLE MAGIC SLATE** (W/S 1963) 8"x12" illustrated cardboard display card holds lift-up erasable film sheet and comes with wood stylus. **$25-50**

12

12) **BULLWINKLE MOTORIZED TARGET GAME** (Parks 1961) 22"x17"window display box contains three-dimensional action scene with three villain targets (Boris Badenov, Snidely Whiplash and Fearless Leader). When one villain is knocked down, another is automatically reset as bell rings. Targets run by wind-up spring. Set includes plastic spring-loaded luger and five rubber-tipped darts. **$100-200**

13) **BULLWINKLE SOAKY** (1963) 12" plastic soap container with hard plastic removable head. **$20-30**

14

15

14) BULLWINKLE SPELLING AND COUNTING BOARD (Laramie 1969) 9"x16"display card holds red plastic plate device with movable letters and numbers to spell and count. **$10-15**

15) BULLWINKLE T-SHIRT (1972) Colorful illustration of Bullwinkle. **$15-20**

16) BULLWINKLE STAMP SET (Laurime 1970) Comes on 10"x6"card **$15-20**

17) BULLWINKLE STUFFED TALKING DOLL (Mattel 1970) 18" brown stuffed plush doll with molded vinyl plastic face and hands has a voice activated pull string which, when pulled, produces eleven different phrases. **$50-75**

18) BULLWINKLE "SUN-EZE" PHOTO DEVELOP SET (1962) 3"x4" box contains materials for developing film negatives on paper by light and producing a "Sun-Eze" print. **$20-25**

19

19) BULLWINKLE SUPERMARKET BOARD GAME (Whitman 1976) **$25-40**

20

20) BULLWINKLE TARGET AND RING TOSS GAME (Parks 1961) Set contains five die-cut, rigid stand-up target figures of Bullwinkle, Rocky, Boris, Natasha and Fearless Leader, luger, four darts and three rings. Comes in 9"x12" package with illustrated header card. **$50-100**

21

22

21) BULLWINKLE TATTOO WRAPPER (Fleer Corp. 1965) 1.5"x3.5"tattoo wrapper features portrait of Bullwinkle on wrapper. Tattoo on reverse side features one of many different Jay Ward characters. **$15-25**

22) BULLWINKLE VENDING GUM BALL DISPLAY CARD (1972) 4"x5"stiff paper card used in gumball machines to promote Bullwinkle tattoos. **$20-25**

23) BULLWINKLE TRAVEL ADVENTURE GAME (TransOgram 1970) 15"x18" box contains playing board, spinner, four playing pieces and 36 "Rocky Saves" cards. Object of the game to move your marker from the home area to the finish area while avoiding pitfalls along the way. **$20-35**

24

26) DUDLEY DO-RIGHT COLORING BOOK (Whitman 1972) 8"x11" 60-pages. **$15-20**

27

28

24) BULLWINKLE TWIN CORK CANNON SET (Parks 1961) 17"x22"window display box contains two plastic cannons which shoot six corks each and are powered by air hose/pump, plus six die-cut rigid stand-up targets of villains (Boris, Natasha, Snidely Whiplash, Fearless Leader, etc.). **$150-200**

27) DUDLEY DORIGHT JIGSAW PUZZLE (Fairchild 1971) Nice colorful scene taken from actual cartoon cell with Dudley, Nellie and Snidely Whiplash. **$15-25**

28) SNIDELY WHIPLASH DRINKING GLASS (Pepsi 1970's) Tall 6" glass. Uncommon. **$15-20**

25

29

29) DUDLEY DO-RIGHT SCHOOL BAG (Ardee Ind. 1972) 8"x12"x3"bright red composition school bag with illustration of Dudley saving Nell from Snidely Whiplash. This scene is from the opening sequence of the cartoon itself. **$50-100**

30

25) BULLWINKLE "FLI-HI" TARGET GAME FEATURING DUDLEY-DO-RIGHT (Parks 1961) 15"x14" window display box contains sturdy die-cut cardboard stand-up target depicting Snidely Whiplash kidnapping Nell on horse as Dudley chases after. The Snidely Whiplash figure is actually a detatchable spring-loaded target and, when hit by dart, will "fli-hi" into the air. Set comes with plastic luger and four rubber-tipped darts. **$150-250**

30) ROCKY & BULLWINKLE MAGIC DOT GAME (Whitman 1962) 15"x8"box contains playing board, spinner, magic crayons and tissue. Object of the game is to be the first player to draw the outlines that finish the dotted pictures of the characters. The one with the most colored pictures wins. **$40-75**

31

31) ROCKY & BULLWINKLE PRESTO "SPARKLE" PAINTING SET (Kenner 1962) Deluxe 18"x12"boxed set contains six tubes of watercolors, five tubes of "jewel-like" colors with glitter mixed in the paint to give the finished painting that "sparkle" look. Set also includes six pre-numbered sketches and three 8"x10"comic strip panels. **$100-150**

32

32) ROCKY & HIS FRIENDS BANK (1961) 7"x5"painted glazed china bank depicting Rocky, Mr. Peabody and Bullwinkle in parade formation carrying flags and signs, which together read "Rocky & His Friends Bank." **$150-250**

33

33) ROCKY & HIS FRIENDS CARTOON KIT (Colorforms 1961) 8"x12"box contains plastic vinyl stick-on character parts which affix to illustrated background. (**Note:** Two diffent contents exist, although the same box lid design is used. One set depicts an outer space theme with background board showing a rocket landing on moonscape and character pieces with space gear. The other set has a landscape background board and non-space equipped character pieces.) **Space set: $100 Non-Space set: $50**

34

35

34) ROCKY & HIS FRIENDS CERAMIC WALL PLAQUE (1961) 10"x7"painted ceramic wall plaque depicting Rocky, Bullwinkle and Mr. Peabody. **$150-200**

35) ROCKY & HIS FRIENDS CHILDREN'S RECORD (Little Golden 1961) 5"x7"illustrated sleeve contains 45 rpm record with two songs. **$15-20**

36

36) ROCKY & HIS FRIENDS GAME (Milton Bradley 1960) 18"x10"box contains playing board, pieces, spinner, 32 character cards and metal school bell. Object of the game is to be the first player to fill eight school desks in the school room with the proper character cards. **$50-75**

37

37) ROCKY & HIS FRIENDS SEWING CARDS (Whitman 1961) Set includes six 6"x8"color illustrated cards with small holes along the outline of the character which can be sewn along and "traced" by yarn. Seldom featured characters like **Captain Wrongway Peach Fuzz** and **Cloyd and Gidney Martians** are included. **$75-100**

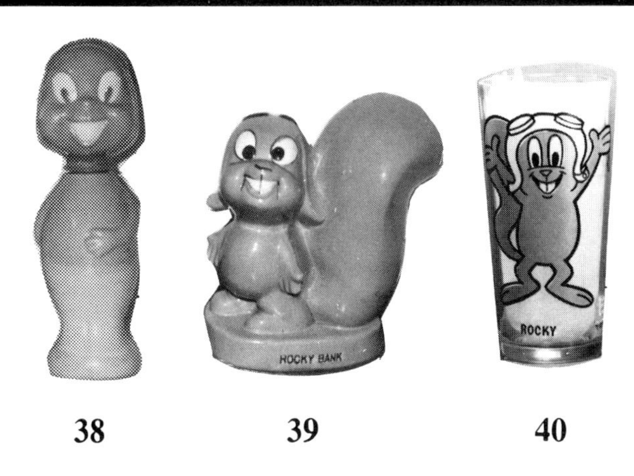

38 39 40

38) **ROCKY SQUIRREL "SOAKY" BUBBLE BATH CONTAINER** (Colgate-Palmolive 1962) 10" plastic soap container with hard plastic removable head. **$15-25**

39) **ROCKY SQUIRREL BANK** (1961) 6" tall painted glazed china figural bank of Rocky with blue cap standing on green base which has decal bearing his name. **$150-250**

40) **ROCKY SQUIRREL DRINKING GLASS** (Pepsi 1970s) Color 6" tall glass. **$10-15**

41 42

41) **ROCKY THE FLYING SQUIRREL COLORING BOOK** (Whitman 1965) 8"x11", 80-pages. **$15-30**

42) **GEORGE OF THE JUNGLE COLORING BOOK** (Whitman 1968) 8"x11", 80+ pages. **$20-30**

43

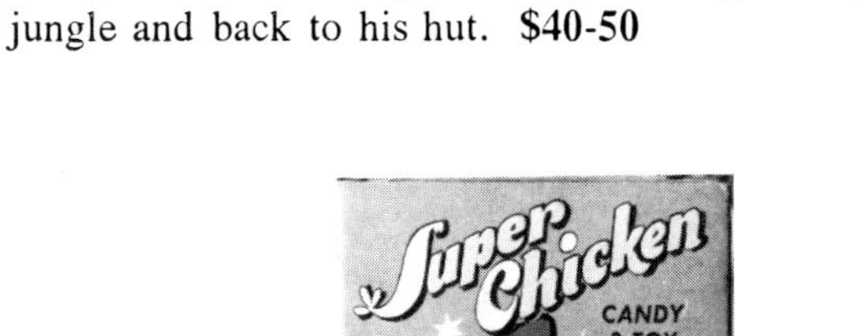

43) **GEORGE OF THE JUNGLE GAME** (Parker Bros. 1968) 10"x19" box. Object is to get George safely through the jungle and back to his hut. **$40-50**

44

44) **SUPER CHICKEN CANDY BOX** (Phoenix Candies 1968) 2.5"x3.5" cardboard box features illustrations of Super Chicken and his sidekick Fred on all six sides. Box originally contained candy and free premium toy. **$25-50**

MR. MAGOO

The first Mr. Magoo cartoon premiered in 1949 as a theater cartoon and continued to entertain the movie audience until 1959. The cartoon series won the Academy Award twice for "Best Short Subject" category in the 1950's. The Magoo series was syndicated to television and in 1960, 130 new five minute color episodes were produced for television. Besides a large toy merchandising campaign, Magoo was General Electric's mascot in the late Fifties through the early Sixties. In October of 1964, "The Famous Adventures of Mr. Magoo" debuted, which featured Magoo portraying famous historical people. The half-hour cartoon series was cancelled in August of 1965.

1

1) MR. MAGOO & WALDO RING (Macman Ent. 1956) Ring comes with two interchangeable heads, one of Mr. Magoo and the other Waldo or Charlie. Comes bagged on 2"x4"illustrated header card. **$20-25**

2) MR. MAGOO BIG TOP BIRTHDAY BASH PREMIUM (GE/Hershey 1970s) Large illustrated envelope contains a variety of circus related items with Magoo as the ringmaster. There are 8 postcard invitations, cut-out animals, circus train Magoo figure, large 36"x24"poster, coloring book and more. **$50-60**

4

4) MR. MAGOO CAR (Hubley 1961) 9" long battery-operated yellow tin litho car with cloth top. Full vinyl figure of Magoo sits behind the wheel. Car moves forward with a shaky, jerking motion. Box is 10"x8"x5"deep. **$250-350**

3

3) MR. MAGOO BOARD GAME (Standard Toykraft 1964) 20"x10"box contains playing board, four red plastic magnifying glasses, 24 magic cards and playing pieces. Object is to move through Magoo-land sharing adventures and misadventures and arriving home safely. **$35-65**

5

5) **MR. MAGOO PLASTIC "STICK-ONS"** (1962) 8"x12" box contains die-cut thin vinyl stick-on character pieces and accessory pieces, plus background scene for placement of characters, and fold-out instruction booklet. **$25-50**

6

7

6) **MR. MAGOO CUT-OUT COLORING BOOK** (Golden 1960) 8"x11",100-pages. **$15-30**

7) **MR. MAGOO PAPERBACK** (Pyramid 1967) Newspaper cartoon reprints. **$8-10**

8

8) **MR. MAGOO PLAYING CARDS** (GE Electric 1960s) Double pack in plastic casing. **$10-20**

9

9) **MR. MAGOO POP UP TARGET SET** (Knickerbacker 1956) Large colorful 22"x20"window display box contains thick cardboard made 20x15" stand-up color target with Waldo, Grandma Magoo and Mr. Magoo, who shoots up out of a cannon if bullseye is hit. Set is entitled "Mr. Magoo at the Circus." Super display piece. **$100-150**

10

11

10) **MR. MAGOO "SAFETY SPIN" 8mm FILM** (Columbia Pictures 1961) Colorful 4"x4"box contains cartoon film. **$5-10**

11) **MR. MAGOO TATTOO WRAPPER** (Fleer 1967) 1.5"x3.5"tattoo wrapper features Magoo tattoo on reverse side. **$8-15**

12

13

12) **MR. MAGOO "SOAKY" BUBBLE BATH CONTAINER** (1963) 12"plastic bubble bath container. **$10-15**

13) **MR. MAGOO STEREO LP "MAGOO IN HI-FI"** (RCA 1956) Voice of Jim Backus. **$10-15**

14

15

14) MR. MAGOO STUFFED DOLL (Ideal 1960) 15" doll with molded vinyl head and cloth stuffed body dressed in jacket, slacks, scarf and hat. **$50-65**

15) MR. MAGOO VINYL FIGURE (1958) Detailed 12" figure in green overcoat with fur collar and hat with walking cane. **$100-200**

16

16) MR. MAGOO "VISITS THE ZOO" GAME (Lowe 1961) 10"x20"box contains cardboard figures of Magoo and zoo animals. Object is to get Magoo safely through the zoo. **$35-65**

17

17) "OH MAGOO, YOU'VE DONE IT AGAIN" BOARD GAME (Warren 1973) **$15-20**

18

18) THE NEARSIGHTED MISTER MAGOO MADDENING ADVENTURES GAME (TransOgram 1970) 18"x9"box. **$20-30**

POPEYE

Popeye is the only cartoon character to have starred in more than 450 episodes. His debut episode appeared before theaters in 1933 for Paramount Pictures, and was created by Fleischer Studios/Famous Studios Productions. In 1961, new episodes were produced especially for television by King Features Productions and the new Popeye sported an all-white sailor suit for the occasion.

1

2

1) **POPEYE AND HIS FRIENDS "3 ON 1" RECORD** (Golden 1962) Illustrated 7"x8"stiff cardboard sleeve contains 33-1/3 record containing six songs including the famous "I'm Popeye the Sailor Man" and "I had a Hamburger Dream." Features three complete records in one and is illustrated on sleeve by showing three individual record sleeves of Popeye, Olive Oyl and Wimpy. **$8-12**

2) **POPEYE AND HIS FRIENDS "3 ON 1" RECORD** (Golden 1961) "Five songs of friendship, safety, health, and manners." Record comes in 7"x8" stiff cardboard sleeve showing three individual record sleeves of Popeye. **$5-10**

3

3) **POPEYE BOARD GAME** (TransOgram 1957) 10"x18" box contains playing board which features four individual Popeye adventure scenes, one spinner, playing pieces, 32 spinach can markers, one spinach shaker box, 24 lottery discs and 16 illustrated adventure cards. Object of the game is to be the first player to win each of the four adventures on the playing board by eating the right number of cans of spinach. **$35-60**

4

4) **POPEYE BUBBLE 'N CLEAN BUBBLE BATH** (The Woolfoam Corp. 1950's)6"x9"x2"thick box features illustration of Popeye and Olive Oyl on front and a live baby turtle offer on back. **$50-75**

5

5) POPEYE BUITONI (Buitoni Food Corp. 1960) 4"x9"box contains green macaroni with added spinach in the shape of Popeye. The back of the box features stand-up cut-outs of Popeye, Brutus, Jeep, Swee' Pea and Wimpy. There is also a free toy inside. The side panel has an exclusive mail-in offer for a 12 inch 33-1/3 rpm Popeye record by RCA. **$75-100**

6

6) POPEYE CARD GAME (Ed-U-Cards 1958) Deck of playing cards with illustrations of Popeye in various action scenes comes in 3"x4"illustrated protective box. **$8-12**

7

8

7) POPEYE CHILDRENS' RECORD (1964) Colorful die-cut cardboard sleeve in the shape of Popeye's head. **$8-12**

8) POPEYE COOKIE JAR (McCoy 1965) 11" tall canister-shaped ceramic cookie jar with color illustration of Popeye, Olive Oyl and Wimpy. **$75-150**

9

9) POPEYE COLORFORMS SET (1957) 8"x12" box contains die-cut thin vinyl stick-on character pieces and accessory pieces, plus background scene for placement of characters, and fold-out instruction booklet. **$35-45**

10

11

10) POPEYE DOLL (Gund 1958) 20" tall stuffed doll with cloth sailor suit and molved vinyl head and arms. **$75-100**

11) POPEYE HALLOWEEN COSTUME (Collegeville 1959) 10"x12"box contains mask and one-piece fabric bodysuit with illustration of Popeye on front. **$25-50**

12

12) POPEYE'S HARMONICA (Plastic Injector Corp. 1958) 5"x5"illustrated display card holds 5" red plastic harmonica. Back of card has three step-by-step harmonica songs. **$25-50**

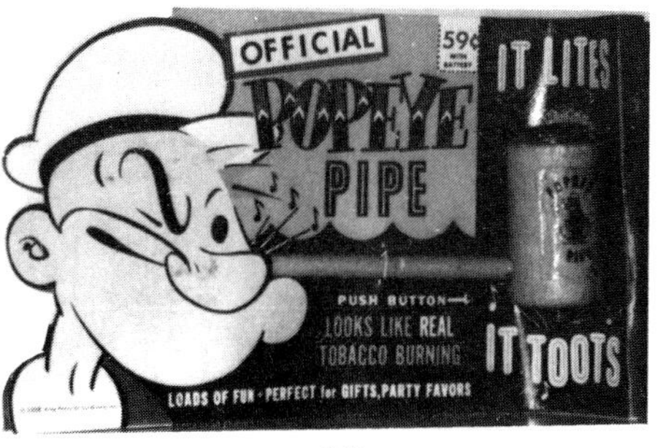

13

13) POPEYE LIGHT UP PIPE (1958) Colorful 9"x7" illustrated box contains an actual pipe that lights up! **$25-50**

14

14) POPEYE LUNCHBOX WITH THERMOS (1964) Metal box: **$45-85** Metal thermos: **$20-40**

15

15) POPEYE TV ERAS-O BOARD SET (Hasbro 1958) Large 18"x8"box contains several illustrated erasable boards, crayons, and wipe cloth. **$50-65**

16

17

16) POPEYE NUMBERED PENCIL COLORING SET (Hasbro 1950) 13"x10"box contains illustrated cardboard inlay which holds twelve color pencils, pencil sharpener, and twelve pre-numbered sketches in two different sizes. **$50-75**

17) POPEYE OIL PAINTING-BY-NUMBER (Hasbro 1958) 7"x10"window display box contains three numbered canvas panels, six oil paints and brush. **$25-50**

18

18) POPEYE PILLOW (1950's) 14"x10" stuffed pillow features large illustration of Popeye and several smaller illustrations of supporting characters. **$15-30**

19

20

19) POPEYE POPCORN (Purity Mills 1949) 4"x5"x3"tin canister contains 12 ox. of white hull-less popcorn. **$25-75**

20) POPEYE PRESTO PAINT SET (Kenner 1961) 7"x9"box contains eight pre-numbered sketches and six tubes of Presto paints. **$25-50**

21

21) POPEYE TALKING HAND PUPPET (Mattel 1967) 18" tall cloth body puppet with hard plastic face has pull string which activates concealed voice box, producing a variety of phrases. Puppet comes with removable wood pipe. Box is 7"x12"x5"deep. **$50-75**

22 23

22) **POPEYE PUSH BUTTON PUPPET** (Kohner 1963) 3" painted, plastic, jointed, mechanical figure puppet which moves in a variety of poses when bottom of base is depressed. **$10-20**

23) **OLIVE OYL BUTTON PUPPET** (Kohner 1963) 3" painted, plastic, jointed, mechanical figure puppet which moves in a variety of poses when bottom of base is depressed. **$10-20**

24

25

24) **POPEYE TATTOOS** (1963) 1.5"x3.5"tattoo wrapper features Popeye on the wrapper side and Bluto and/or Popeye on the tattoo side. Two different style wrappers.

> **A) Popeye, first series** half-figure against blue background
>
> **B) Popeye, second series** Popeye face against yellow background
>
> EACH: **$25-40**

25) **POPEYE VINYL BANK** (Alan Jay 1959) 8" tall molded vinyl figure depicts Popeye sitting down with crossed legs holding a can of spinach. The coin slot is on top of the can of spinach. Figure comes in color accents with blue pants and cap and black shirt with red collar. **$50-75**

26

27

26) **POPEYE VIEWMASTER REEL SET** (Sawyer 1962) 5"x5"color illustrated envelope contains three reels and 16-page story booklet. **$25-50**

27) **POPEYE TRAY PUZZLE** (Jaymar 1962) Inlay jigsaw puzzle in a 11"x14"frame tray. **$10-12**

28

29

28) **POPEYE THE WEATHERMAN COLORFORMS SET** (1957) 8"x12"box contains die-cut thin vinyl stick-on character pieces and accessory pieces, plus background scene for placement of characters, and fold-out instruction booklet. **$50-65**

29) **TV COLORING BOOK FEATURING POPEYE THE SAILOR** (Lowe 1961) 8"x11", 80+ pages. **$12-15**

30

30) **THIMBLE THEATER GAME STARRING OLIVE OYL IN "I'LL CATCH MY POPEYE"** (Hasbro 1965) 18"x9"box contains playing board, playing pieces and spinner. **$25-35**

TERRYTOON CHARACTERS

Creator/animator Paul Terry formed Terrytoons in the early 1940's with his most famous character, Mighty Mouse. Other notable characters include Heckle and Jeckle, Dinky Duck, Gandy Goose, and Little Roquefort, which remained strong sellers in film and toy merchandising throughout the Forties and Fifties. The early Sixties saw new cartoon characters created for television which included Deputy Dawg, Sidney the Elephant, Hashimoto-San and the Mighty Heroes. Hector Heathcote, which premiered in its own half-hour show on NBC from 1963-64, was actually created in 1959 and originally appeared as a theatre cartoon prior to syndication.

1

1) TERRYTOON FRAME TRAY PUZZLES (Jaymar 1950's) Inlay jigsaw puzzles in a 11"x14"frame tray featuring a variety of Terrytoon characters including Mighty Mouse, Heckle and Jeckle, Gandy Goose and Tom Terrific. EACH: **$10-20**

2

2) TERRYTOON FUN KIT (MB 1958) 5"x8"window display box contains crayons, color pages of Terrytoon characters and colorful cut-out figures. **$15-25**

3

3) TERRYTOON'S "HIDE N' SEEK" GAME (Transogram 1960) 9"x17" box features several Terrytoon characters including **Tom Terrific, Dinky Duck, Heckle & Jeckle** and more. Comes with die-cut cardboard character heads, spinner and board. **$35-65**

4

4) **CARTOONARAMA CARTOON CEL PAINT SET** (Cartoonarama 1970) 14"x10"boxed set contains six 9"x10" illustrated cartoon cells, each featuring a different Terrytoon character (Deputy Dawg, Mighty Mouse, Astronaut, Tom Terrific, Stanley the Lion, Sidney the Elephant), six picture frames and eight vials of paint and brush. Box reads "Actual process used in animated cartoon." **$35-65**

5

5) **DEPUTY DAWG "ACTION PUSH-OUT" FIGURES** (Lays Potato Chips) Colorful 8"x10"cardboard sheets contain punch-out stand-up figures of Deputy Dawg and friends plus action dioramas of henhouse, creek mud, jail house, etc., and a variety of games and puzzles. Ten different sheets were made, one available in every snack-size six pack of Lays potato chips. EACH: **$10-15**

6

7

6) **DEPUTY DAWG BADGE** (Laramie 1973) Metal badge on 4"x6"display card. **$8-10**

7) **DEPUTY DAWG BOP BAG** (Doughboy 1961) 54" tall inflatable vinyl plastic bounce-back bop bag with weighted, rolled bottom. **$40-50**

8

8) **DEPUTY DAWG BOARD GAME** (MB 1960) 20"x10"box contains three decks of playing cards, playing pieces and board. Object of the game is to overcome obstacles and be the first player home. **$25-45**

9) **DEPUTY DAWG COLORING BOOK** (Treasure 1961) 8"x11",60+ pages. **$15-25**

10

11

10) **DEPUTY DAWG HALLOWEEN COSTUME** (B.C. 1961) 10"x12"box contains mask and one-piece fabric bodysuit with illustration of Deputy Dawg on front. **$30-50**

11) **DEPUTY DAWG MAGIC SLATE** (1962) 8"x12" illustrated cardboard display card holds lift-up erasable film sheet and comes with wood stylus. **$25-30**

12

12) **DEPUTY DAWG PENCIL BOX** (Hasbro 1961) 4"x8"x1" red cardboard box with snap open lid and pull-out drawer. Illustrated colored paper decal of characters on top lid. **$15-25**

13

14

13) **DEPUTY DAWG RECORD ALBUM** (RCA-Camden 1961) Theme song and stories with original TV soundtrack voices. **$10-20**

14) **DEPUTY DAWG SOAKY BUBBLE BATH CONTAINER** (1963) 10" plastic soap container with hard plastic removable head. **$15-20**

15

15) **DEPUTY DAWG STORY BOOKS** (Wonder Books 1961) 7"x8"hardback book with 20+ page story with full color story art on each page. EACH: **$10-15**

16) **DEPUTY DAWG STUFFED DOLL** (Ideal 1961) 16" stuffed doll with hat, badge and gun. **$75-100**

17

19

17) **DEPUTY DAWG VIEWMASTER REEL SET** (Sawyer 1961) 5"x5" illustrated envelope contains three viewmaster reels and 16-page booklet. **$10-20**

18) **MUSKY MUSKRAT "SOAKY" BUBBLE BATH CONTAINER** (Colgate-Palmolive 1963) 10" plastic soap container with hard plastic removable head. **$15-25**

19) **DINKY DUCK MAGIC SLATE** (Lowe 1952) 8"x12" illustrated cardboard display card holds lift-up erasable film sheet and comes with wood drawing stylus. **$10-15**

20

20) **DINKY THE DUCK TRAY PUZZLES** (Fairchild 1951) Inlay jigsaw puzzle in a 11"x14"frame tray. EACH: **$15-20**

21

21) **HASHIMOTO-SAN GAME** (Transogram 1963) 10"x17" game. Object is to be the first player who successfully ties down Dangerous Cat. **$25-35**

22

22) **HASHIMOTO-SAN PAINT BY NUMBER COLORING SET** (TransOgram 1965) 16"x9"window display box contains plastic paint tray with eight inlaid paint tablets, brush and eight pre-numbered sketches. **$40-60**

23 24

23) **HECKLE & JECKLE CHILDREN'S RECORD** (Little Golden 1958) 7"x8" stiff paper sleeve containing 45 rpm children's record. **$8-12**

24) **HECKLE & JECKLE COLORING BOOK** (Treasure 1957) 8"x11",60+ pages. **$10-15**

25 26

25) **HECKLE & JECKLE FIGURE** (1958) 8" hard rubber figure. **$50-75**

26) **HECKLE & JECKLE MAGIC SLATE** (Lowe 1952) Die-cut magic slate featuring both Magpies. **$15-25**

27) **HECKLE & JECKLE 3-D TARGET GAME** (TV Film Sales 1958) Colorful 14"x10"plastic target set with characters super embossed to create a 3-D image. **$25-30**

28

28) **HECTOR HEATHCOTE CARTOON KIT** (Colorforms 1964) 8"x12"box contains plastic vinyl character parts that affix to illustrated background board. **$35-50**

29

29) **HECTOR HEATHCOTE LUNCHBOX WITH THERMOS** (1963) Box: **$125-175** Thermos: **$35-75**

30

32

33

33) **HECTOR HEATHCOTE SHOW RECORD ALBUM** (RCA-Camden 1963) Contains theme song and six short stories featuring Hector Heathcote, Hashimoto-San and Sidney the Elephant in original TV soundtrack voices. **$25-35**

30) **HECTOR HEATHCOTE "THE MINUTE-AND-A-HALF MAN" GAME** (Transogram 1963) Object of the game is to take Hector safely through the Revolutionary War and receive the most medals. 10"x20"box contains four Hector figures, 20 battle cards, 44 medal cards, cannonade and marble. **$65-100**

34

35

34) **HECTOR HEATHCOTE WONDER BOOK** (1962) 7"x8"hardback book with 28-page story accompanied by full color story art on each page. **$10-12**

35) **MIGHTY MOUSE ALARM CLOCK** (1960's) 5" diameter metal alarm clock with full color image of Mighty Mouse on face and his arms as the hour and minute hands. **$35-75**

31

31) **HECTOR HEATHCOTÉ PENCIL BY NUMBER COLORING SET** (Transogram 1964) 10"x17"window display box contains 12 pre-numbered sketches, six pencils and pencil sharpener. **$50-75**

32) **HECTOR HEATHCOTE SHOW MAGIC SLATE** (Lowe 1964) 8"x12"cardboard card holds lift-up erasable film sheet and comes with wood drawing stylus. Card features Hector Heathcote, Hashimoto-San and Sidney the Elephant. **$15-25**

36

36) MIGHTY MOUSE CARTOON DRAWING SET (Gabriel 1956) 18"x12"boxed cartoon set contains 28-page "How to Draw Funny Cartoons" book, a Mighty Mouse sketch pad, crayons and pencil. Illustrated throughout. **$40-65**

37

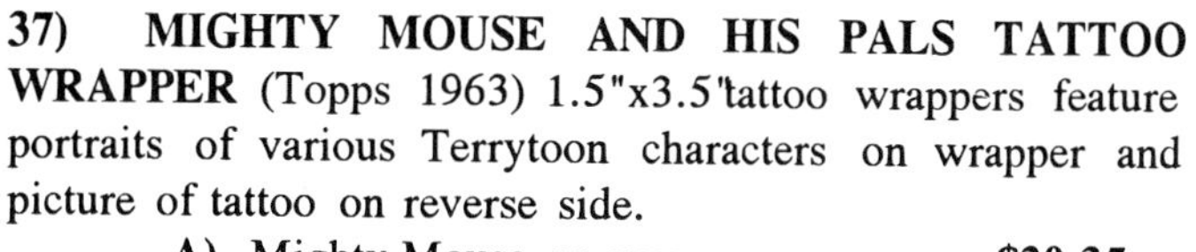

37) MIGHTY MOUSE AND HIS PALS TATTOO WRAPPER (Topps 1963) 1.5"x3.5"tattoo wrappers feature portraits of various Terrytoon characters on wrapper and picture of tattoo on reverse side.

A) Mighty Mouse on wrapper	**$20-35**
B) Heckle and Jeckle on wrapper	**$20-35**
C) Deputy Dog on wrapper	**$35-50**

38

39

38) MIGHTY MOUSE CHILDREN'S RECORD (Little Golden 1958) 7"x8"illustrated stiff paper sleeve containing 45 rpm children's record with cartoon theme song. **$10-15**

39) MIGHTY MOUSE COLORING BOOK (Treasure 1957) 8"x11",60-page illustrated coloring book with story. **$10-15**

40

40) MIGHTY MOUSE FRAME TRAY PUZZLES (Jaymar 1950's) Inlay jigsaw puzzles in a 11"x14"frame tray. EACH: **$8-12**

41

42

41) MIGHTY MOUSE GAME (MB 1958) 5"x8"window display box contains illustrated target and tiddly winks. **$10-20**

42) MIGHTY MOUSE JIGSAW PUZZLE (Whitman 1967) 8"x11"box contains colorfully illustrated 100-piece puzzle of mighty mouse saving a bridge from collapsing. Puzzle is 14"x18"when assembled. **$10-15**

43

43) MIGHTY MOUSE MAGIC SLATES (Lowe 1950s) 8"x12"illustrated cardboard display card holds lift-up erasable film sheet and comes with wood stylus. EACH: **$20-25**

44

45

50

51

44) MIGHTY MOUSE PRESTO PAINT SET (Kenner 1963) Bright attractive 8"x10" boxed set also has featured Heckle & Jeckle and Deputy Dawg. **$25-50**

45) MIGHTY MOUSE RESCUE GAME (HG 1956) Colorful game with cardboard stand-up figures of the Terrytoon characters included! **$35-45**

46) MIGHTY MOUSE SOAKY BUBBLE BATH CONTAINER (1963) 10" plastic soap container with hard plastic removable small head. **$15-20**

47) MIGHTY MOUSE SWIMMING INNER-TUBE (Ideal 1956) Colorful 18"x15"plastic vinyl inflatable inner-tube with graphics of **Mighty Mouse, Gandy Goose, Roquefort Mouse, Dinky Duck and Pearl Pureheart. $50-65**

48

49

48) MIGHTY MOUSE & HIS PALS GAME (MB 1957) 13"x13" box contains very colorful 24"x13" playing board showing Mighty Mouse and 16 other Terrytoon characters including Tom Terrific, Heckle & Jeckle and Dinky Duck. Game includes spinner and playing pieces. Object is to be the first player to get all his playing pieces safely home. **$35-60**

49) MIGHTY MOUSE VIEWMASTER REEL SET (Sawyer 1956) 5"x5"color illustrated envelope contains three reels and 16-page story booklet. **$15-25**

50) MIGHTY MOUSE VINYL SQUEEZE FIGURE (1950's) 10" soft rubber squeeze figure depicts Mighty Mouse in a yellow uniform with red trunks and red cloth cape with the words "Mighty Mouse" across his chest. **$50-65**

51) THE ADVENTURES OF MIGHTY MOUSE RECORD ALBUM (Rocking Horse 1957) Theme song and stories with original TV soundtrack voices. **$8-12**

52

52) SILLY SIDNEY ABSENT-MINDED ELEPHANT BOARD GAME (TransOgram 1963) 15"x8" box contains playing board, playing pieces, spinner and 16 illustrated invitation cards. Object of the game is to be the first player to visit all the animals from whom he receives invitations. **$50-75**

53

53) SILLY SIDNEY "COCONUT SHOOT" TARGET GAME (TransOgram 1963) 19"x15" window display box contains 16" long plastic gun molded in the shape of Silly Sidney the Elephant, two hollow plastic figures of the owl and lion, and three "coconut balls". When elephant is squeezed, coconuts shoot from his trunk. **$100-150**

54

55

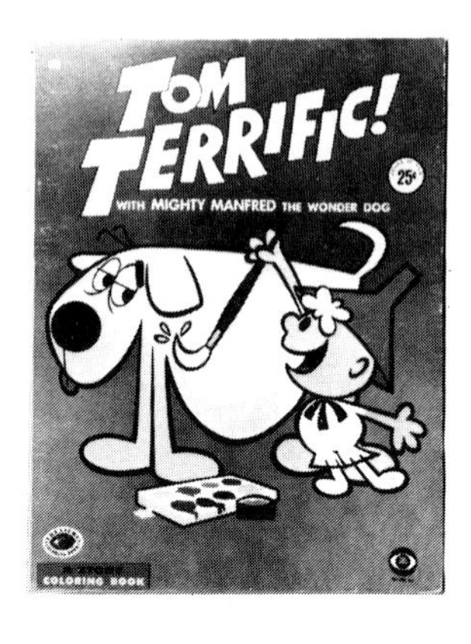

56

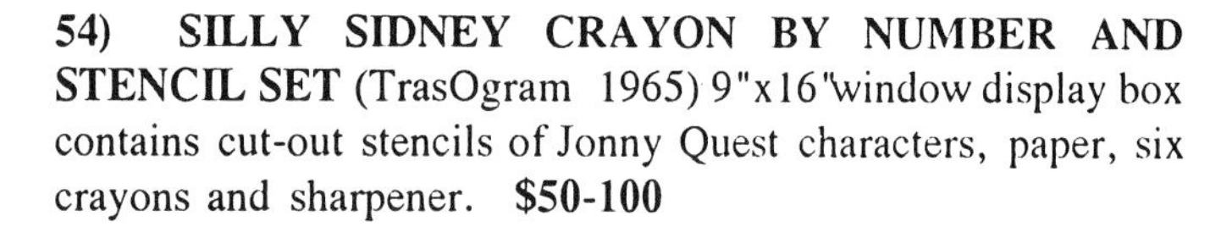

54) SILLY SIDNEY CRAYON BY NUMBER AND STENCIL SET (TrasOgram 1965) 9"x16" window display box contains cut-out stencils of Jonny Quest characters, paper, six crayons and sharpener. **$50-100**

55) TOM TERRIFIC BOOK (Wonder 1960) 7"x8" hardback book contains 20+ page story with full color story art. **$8-15**

56) TOM TERRIFIC--A STORY COLORING BOOK (Treasure Books 1957) 8"x11", 50-pages. **$20-25**

WALTER LANTZ CHARACTERS

Walter Lantz's animation talents can be traced back to 1915 when he and producer J.R. Bray introduced the cartoon short "Colonel Heeza Liar," through Hod-Kinson and Selznick Pictures. In 1939, Lantz created his own character, Andy Panda (released through Universal Pictures), and met with such success that he was contracted to produce four to five more Andy Panda cartoons a year. Woody Woodpecker made his debut 1940 in the Andy Panda cartoon "Knock Knock". Woody was created by Lantz on his honeymoon when a non-stop pecking woodpecker bore a hole through their cottage roof, causing many a restless night! Woody has since become the celebrated icon of the Lantz cartoon clan. The Woody Woodpecker show premiered on television in 1957 as a half-hour cartoon show on ABC. NBC later re-syndicated the program in 1970 to 1972. Lantz continued doing theatrical cartoon shorts for theaters until 1972 and was the last of his breed for this media of entertainment.

1

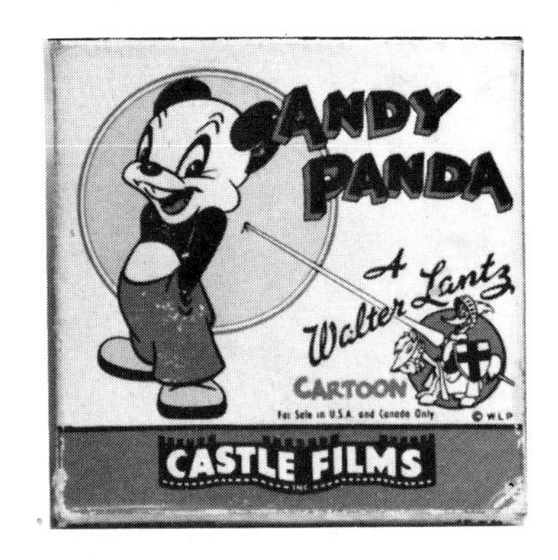

2

3

4

1) **ANDY PANDA BANK** (1948) 6" tall pressed cardboard painted figure. **$18-25**

2) **ANDY PANDA CARTOON FILM** (1940's) 16mm film comes in 5"x5"illustrated box. **$5-12**

3) **THE BEAR FAMILY ALBUM - A COLORING BOOK** (Saalfied/Artcraft 1964) 8"x11",80-pages. **$10-15**

4) **CHILLY WILLY SQUEEZE FIGURE** (1950's) 6" rubber squeeze figure. **$20-25**

5

5) CHILLY WILLY & SPACE MOUSE & FRIENDS TRAY PUZZLE SET (Saalfield 1963) Boxed set of four 12"x10"super colorful tray puzzles of Chilly, Space Mouse, Gabby Gator and Wally Walrus comes in 10"x12"window display box. **$40-75**

6

7

6) OSWALD THE RABBIT COLORING BOOK (Saalfield 1950's) 8"x11",30-page. **$10-15**

7) OSWALD THE RABBIT CUT-OUT COLORING BOOK (Pocket Books 1952) 30-pages, 8"x11". **$10-20**

8

9

8) OSWALD THE RABBIT CERAMIC FIGURE (1950's) 4" tall hollow painted ceramic figure depicting Oswald as an umpire on a baseball team in a "Strike one!" pose. **$35-50**

9) WALLY WALRUS CERAMIC FIGURE (1950's) 4" tall hollow ceramic painted figure depicting Wally as a catcher on a baseball team. **$35-50**

10

10) SPACE MOUSE COLORING BOOKS (Saalfield 1963) 8"x11",80-pages. EACH: **$10-15**

11

12

11) WALTER LANTZ AND FRIENDS MAGIC SLATE (1962) 8"x12"illustrated cardboard display card holds lift-up erasable film sheet and comes with wood stylus. **$15-25**

12) WALTER LANTZ "EASY WAY TO DRAW" BOOKS (Whitman 1958) 8"x11",100-pages. Forward by Walter Lantz who shares drawing techniques. Also includes puzzles, activities and games.

Hardback	**$15-25**
Softback	**$10-15**

13

13) **WALTER LANTZ PAPER PARTY PLATE** (1959) Colorful square 10"x10"heavy paper plate featuring lots of Lantz characters (many early obscure ones). **$8-12**

14

14) **WALTER LANTZ PICTURE DOMINOES** (Saalfield 1963) 18"x9"box contains several illustrated cardboard tiles, each depicting a Lantz character. **$25-40**

15

16

15) WOODY.WOODPECKER'S ART GALLERY (Saalfield 1962) 10"x14"box contains several paint-by-number pictures in various sizes, brush and two paint palettes. **$20-40**

16) **WOODY WOODPECKER AND HIS FRIENDS CHILDREN'S RECORD ALBUM** (Cricket Records 1962) 33-1/3 rpm record album contains songs and stories. **$10-20**

17

17) **WOODY WOODPECKER BOARD GAME** (MB 1958) **$20-30**

18

18) **WOODY WOODPECKER CARD GAME** (Fairchild 1958) Small illustrated deck of children's playing cards. Comes in 3"x4"protective cardboard box. **$10-15**

19

19) **WOODY WOODPECKER COLORING BOOKS** (Whitman 1956) 8"x11",80-pages. EACH: **$8-12**

20

20) WOODY WOODPECKER'S CRAZY MIXED-UP COLOR FACTORY GAME (Whitman 1972) 17"x12"box contains rotating playing board with color factory wheel, 48 paint chips, Woody figure and spinner. Object is to be the first player to help Woody to align his color wheel in proper order. **$20-35**

21

22

21) WOODY WOODPECKER TATTOO WRAPPER (Topps 1960) 1.5"x3.5"tattoo wrapper features portrait of Woody Woodpecker on wrapper. Tattoo on reverse side features one of many different Walter Lantz characters. **$20-30**

22) WOODY WOODPECKER TV COLORING PENCIL SET (Conn. Pencil Co. 1958) Colorful 9"x8"box features a TV-shaped display window where an illustrated card with poses of Woody can be pulled from slot on side of box to create a TV-cartoon picture. Set includes four color-by-number pictures, six colored pencils and pencil sharpener. **$20-25**

23

23) FLANNEL BOARD PLAY WITH WOODY WOODPECKER AND HIS FRIENDS (Saalfield 1962) 12"x10"box contains illustrated flannel board and several die-cut flannel characters. Characters affix to background and can be lifted up and re-applied to create several different scene scenarios. **$25-40**

24

24) WOODY WOODPECKER "TRAVEL WITH WOODY" GAME (Cadeco 1956) 12"x19" box which holds a self-contained game board which features a rotating notched flight dial and comes with four playing pieces, 20 pegs and spinner. The object of the game is to be the first player to mail a postcard in five different countries and return to the USA. **$50-75**

25

26

25) SUGAR POPS CEREAL WITH WOODY WOODPECKER (Kelloggs 1968) Canadian distributed Kelloggs Sugar Pop featured Woody Woodpecker as mascot on front of box. **$200-300**

26) WOODY WOODPECKER SLIPPERS (1957) Infant's slippers. **$15-20**

WARNER BROTHERS CHARACTERS

Ever since Bugs Bunny first appeared in the 1938 Porky Pig cartoon "Porky'sHare Hunt", he has become an institution in Warner Brothers cartoon history. The supporting trilogy of stars that evolved around Bugs Bunny and Porky Pig grew, as did their individual popularity, and were enthusiastically merchandised in every conceivable form. The theatrical cartoons reigned from 1938-1963 when Warner Brothers' animation department finally closed, but the cartoons have been in syndication for television ever since.

New cartoons were produced for the television debut of "The Bugs Bunny Show" which premiered October 1960, and could be viewed until 1962 on prime time networks for one half-hour. Other Warner Brothers cartoon television shows include: The Bugs Bunny/Road Runner Hour (1968-71), The Porky Pig Show (1964-65), and the Road Runner Show (1967-68, 1971-72)

1

2

1) **WARNER BROS. HANDPUPPETS** (1950's) 10" tall handpuppets with molded vinyl heads and patterned cloth bodies designed in stripes, polka-dots, etc. (R to L: Bugs Bunny, Porky Pig, Sylvester Cat, Foghorn Leghorn and Elmer Fudd) EACH: **$15-25**

2) LOONEY TUNES TV LUNCHBOX (American Thermos 1959) Metal box depicting a television set with Looney Tune characters on the screen. Matching thermos also used in the Porky's Lunchwagon lunchbox set. Box: **$75-125** Thermos: **$25-50**

3) BUGS BUNNY CHILD'S RECORD (1952) 45 rpm record in paper sleeve. **$5-10**

4

4) **BUGS BUNNY AND FRIENDS CARTOON KIT** (Colorforms 1962) 8"x12"box contains die-cut thin vinyl stick-on character pieces and accessory pieces, plus background scene for placement of characters, and fold-out instruction booklet. **$35-45**

5

7) **BUGS BUNNY GIVE-A-SHOW PROJECTOR** (Kenner/Chad Valley 1968) British subsidiary of Kenner products, Chad Valley released an all Warner Bros. Give-A-Show Projector Set which comes in 11"x18"box and contains a plastic battery-powered slide film projector and 16 color slide strips all featuring different Warner Brothers characters. **$35-50**

8

9

5) **BUGS BUNNY CARTOON KIT** (Colorforms 1958) 8"x12" box contains die-cut thin vinyl stick-on character pieces and accessory pieces, plus background scene for placement of characters, and fold-out instruction booklet. **$45-65**

6

8) **BUGS BUNNY HALLOWEEN COSTUME** (Ben Cooper 1950) Illustrated synthetic one-piece suit depicts Bugs running. Comes with colorful mask in 10"x12"window box. **$15-25**

9) **BUGS BUNNY HAND PUPPET** (circa 1950's) 9"puppet with cloth body and soft vinyl head. **$15-20**

6) **BUGS BUNNY COLORING BOOKS** (1950's - 1960's) The two main coloring book companies which produced Bugs Bunny coloring books were Whitman and Watkins/Strathmore Co. during the 1950's and 1960's. Each is 8"x11" and approximately 60 to 80 pages in length. EACH: **$10-15**

10

7

10) **BUGS BUNNY JACK-IN-THE-BOX** (Mattel 1962) 5"x5"x6"color litho metal crank-wound music box contains pop-up Bugs figure with plastic head and cloth body. **$45-75**

11) **BUGS BUNNY MEETS HIAWATHA CHILDREN'S RECORD** (Capitol 1958) 78 rpm story record featuring the voice of Mel Blanc comes in 10"x10"paper sleeve. **$10-20**

11

12

12) **BUGS BUNNY METAL BANK** (1940's) Well detailed 5" figure of Bugs leaning against a tree eating a carrot. All metal. **$50-100**

13) **BUGS BUNNY SOAKY BUBBLE BATH CONTAINER** (1963) 10" plastic bubble bath figure container. **$12-15**

16

17

16) **BUGS BUNNY TALKING VIEW-MASTER REELS** (GAF 1969) 8"x8" boxed set contains reels and accessories. **$10-12**

17) **BUGS BUNNY TATTOO** (Topps 1971) 2"x4" wrapper pack contains tattoo sheet and slab of gum. Wrapper features Bugs Bunny holding tattoo sheet. **$15-25**

15

14

14) **BUGS BUNNY TALKING DOLL** (Mattel 1962) 27" tall grey/white stuffed plush doll with molded vinyl face and hands has voice-activated pull-string which, when pulled, produces eleven different phrases. **$75-100**

15) **BUGS BUNNY TALKING HAND PUPPET** (Mattel 1962) 18" tall grey/white stuffed plush doll with molded vinyl face and hands has voice-activated pull-string which, when pulled, produces eleven different phrases. **$45-75**

18

18) **BUGS BUNNY FRAME TRAY PUZZLE** (Whitman 1958) Inlay jigsaw puzzle in a 11"x14" frame tray. **$10-20**

19

19) **KOOL-AID FLAVOR PACKS FEATURING BUGS BUNNY** (Kool-Aid 1960's) 3"x4" illustrated stiff paper packets feature Bugs Bunny as Kool-Aid's mascot. EACH: **$12-20**

20

21

20) ELMER FUDD COLORING BOOK (Watkins-Strathmore 1964) 8"x11",100-pages. **$8-15**

21) ELMER FUDD SOAKY BUBBLE BATH CONTAINER (Colgate-Palmolive 1963) 9"figural bubble bath container with removable hard plastic head. **$10-15**

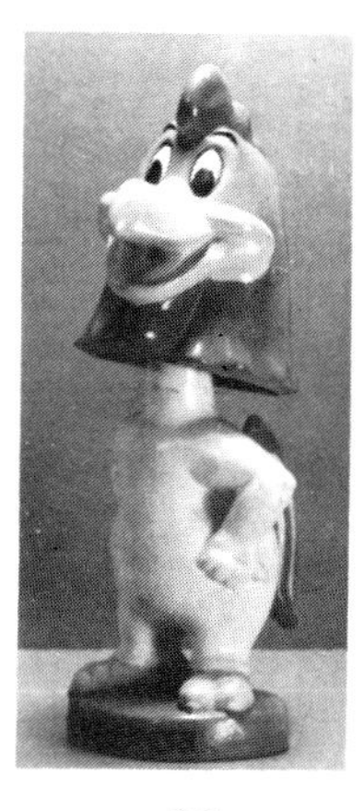

22

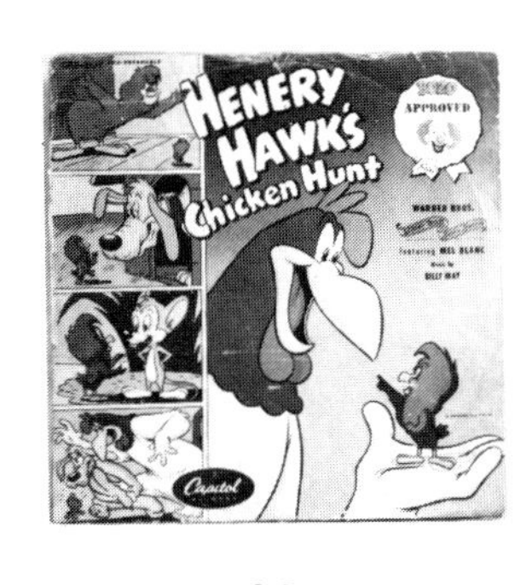

23

22) FOGHORN LEGHORN CERAMIC BOBBING HEAD (Lego 1961) 8" tall painted ceramic nodder. **$50-100**

23) HENRY HAWK'S CHICKEN HUNT CHILDREN'S RECORD (Capitol 1950) 78 rpm story record featuring the voice of Mel Blanc comes in 10"x10"paper sleeve. **$15-25**

24

24) LOONEY TUNES CARTOON-O-GRAPH (1944) Large 20"x14" boxed set - contains colorful drawing board with mechanical arm which allows the user to draw big pictures from smaller ones. Comes w/ several illustrations of Warner Bros. cartoon characters. **$45-75**

25

25) LOONEY TUNES GREETING CARDS w/RECORD (Buzza 1962) Cute colorful set of 6"x6"greeting cards with illustrated record inside extending the greeting (voices by Mel Blanc). EACH: **$8-12**

26

26) LOONEY TUNES "NOTCH-EMS" FIGURES (Amusing Educational 1950) Colorful 8"x11"illustrated box contains eight heavy cardboard figures that are assembled into stand-up figures and include Bugs, Porky, Tweety Bird, Sylvester, Chicken Hawk, Sniffles, Elmer Fudd, and Daffy Duck. Unique and very displayable. **$25-50**

27

27) LOONEY TUNES TOON-A-VISION (1950's) TV-style board with four control knobs allows the user to interchange characters' faces to create over 64,000 faces. **$20-35**

28

28) LOONEY TUNES TRAY PUZZLE (1940's) Inlay jigsaw puzzle in a 11"x14" frame tray featuring Bugs, Elmer and Sylvester. **$15-25**

29

30

29) PORKY PIG AND BUGS BUNNY COLORING BOOK (Watkins-Strathmore 1964) 8"x11",100-pages. **$8-15**

30) PORKY PIG SOAKY BUBBLE BATH CONTAINER (1963) 10" plastic figure bath container. **$10-20**

31) PORKY PIG TALKING HAND PUPPET (Mattel 1964) 18" tall stuffed plush doll with molded vinyl face and hands has voice-activated pull-string which, when pulled, produces eleven different phrases. **$45-75**

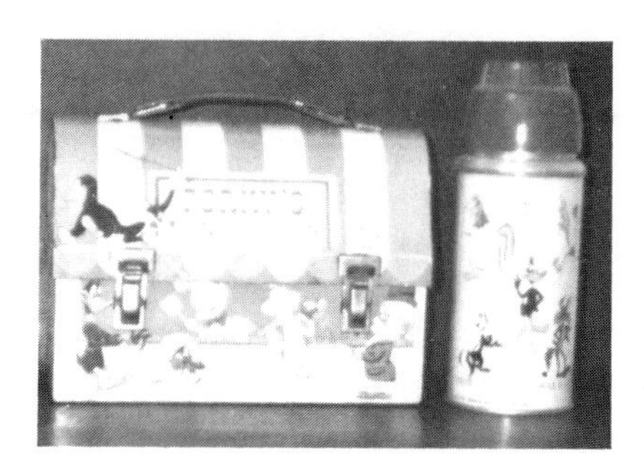

32

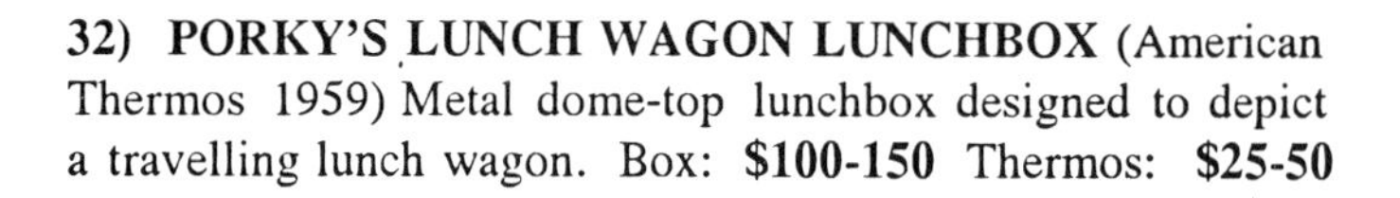

32) PORKY'S LUNCH WAGON LUNCHBOX (American Thermos 1959) Metal dome-top lunchbox designed to depict a travelling lunch wagon. Box: **$100-150** Thermos: **$25-50**

33

33) ROAD RUNNER BOARD GAME (MB 1968) 10"x20" box contains playing board, character cards, illustrated character tiles, playing pieces and spinner. Object of the game is to take the Road Runner and collect four different character cards. **$20-35**

34

35

34) ROAD RUNNER LUNCHBOX (1970) Box: **$50-75** Thermos: **$20-30**

35) BEEP-BEEP THE ROAD RUNNER VIEWMASTER REEL SET (GAF 1960's) Puppet animation scenes comes with booklet in photo envelope. **$12-15**

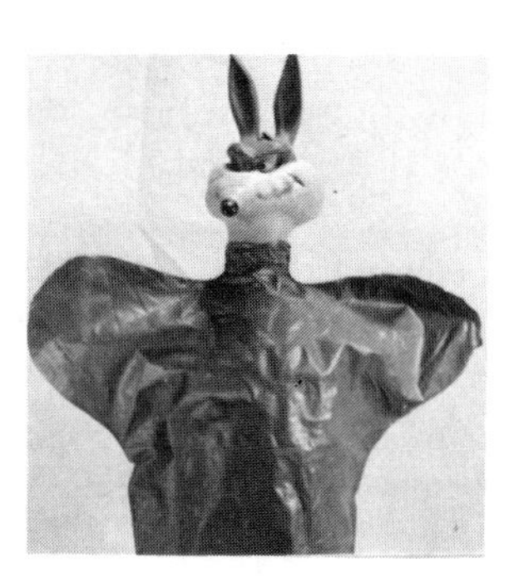

36

36) WILE E. COYOTE HANDPUPPET (1969) 8" handpuppet with brown thin vinyl body and light brown molded vinyl head with color accents. **$15-25**

37

37) SNIFFLES THE MOUSE METAL BANK (1940's) Well detailed 5" metal figure of Sniffles the Mouse standing beside a barrel. All metal. **$75-150**

38

38) SNIFFLES THE MOUSE SQUEEZE FIGURE (Oak Rubber 1951) 6" nicely detailed and painted rubber squeeze figure of the Warner Brothers character "Sniffles." Figure squeaks and comes in 7"x4"x3" window display box with cut-out figures on the side of the box of Sniffles and Mary Jane. **$100-150**

39

40

39) SNIFFLES THE MOUSE STUFFED PILLOW (1960's) 15" tall stuffed pillow doll in shape of Sniffles. **$15-25**

40) SPEEDY GONZALES DAKIN FIGURE (Dakin 1968) Plastic 7" detailed movable figure. **$20-30**

41

42

41) SYLVESTER THE CAT FIGURE (Dakin 1968) 8" hard vinyl plastic figure with movable arms. **$12-18**

42) DANGEROUS DESERT COLORING BOOK FEATURING YOSEMITE SAM (Watkins-Strathmore 1963) 8"x11", 30+ pages. **$8-12**

43

43) YOSEMITE SAM LUNCHBOX (King Seeley Thermos 1971) Vinyl box with matching thermos. Box: **$50-75** Thermos: **$25-50**

MISCELLANEOUS CARTOON CHARACTERS

ARCHIE

A popular comic book character since the early 1940's, Archie made his cartoon debut on CBS in September of 1968. Aided by pop songs which went on to be number one hits, Archie and gang formed "The Archies" and became one of the top cartoons of the late Sixties, spawning five other half-hour Archie cartoon series from 1968 to 1977.

1

1) **ARCHIES FUN GAME** (Hasbro 1963) 10"x20" box contains playing board, spinner, four die-cut stand-up figures of Archie and 20 illustrated playing cards. Object of the game is to get from Riverdale High School to Veronica's swimming pool. **$45-75**

2

2) **ARCHIES GAME** (Whitman 1969) 12"x14"box. **$20-25**

3

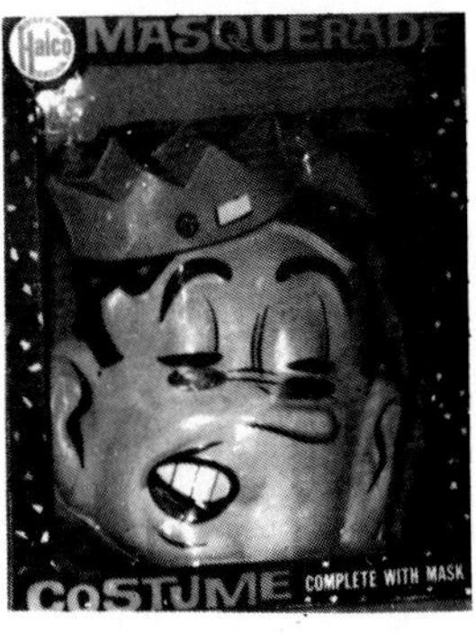

4

3) **ARCHIE HALLOWEEN COSTUME** (Collegeville 1960) 10"x12"box contains mask and one-piece fabric bodysuit with illustration of Archie on front. **$35-50**

4) **JUGHEAD HALLOWEEN COSTUME** (Collegeville 1960) 10"x12"box contains mask and one-piece fabric bodysuit with illustration of Jughead on front. **$50-75**

5

5) **ARCHIE'S JALOPY MODEL KIT** (Aurora 1968) 7"x5" box contains all-plastic assembly kit of jalopy plus figures of Archie and Veronica. **$50-100**

6

7

6) **ARCHIE JIGSAW PUZZLE** (Jaymar 1962) "Swinging Malt Shop". Comes in 7"x10"box. **$15-25**

7) **ARCHIE JIGSAW PUZZLE "THE ARCHIES"** (Whitman 1970) Large 15"x15"box. **$15-25**

8) **ARCHIE'S LUNCHBOX WITH THERMOS** (1968) Steel box: **$50-75** Thermos: **$15-20**

9) **ARCHIE STUFFED DOLL PREMIUM** (1970) 18" colorful Archie doll comes with free comic book. **$15-35**

10

11

10) **ARCHIE'S TATTOOS** (Topps 1969) 2"x4"wrapper pack contains tattoo sheet and slab of gum. Colorful wrapper depicts Archie holding tattoo sheet and covered with tattoos. **$15-20**

11) **ARCHIE'S TOTE BAG** (Fairmont Potato Chips 1971) Large 15"x18"heavy cloth tote bag with drawstring features portrait of Archie and Archie logo and reads "Swing With Archie." Reverse side is designed like a Fairmont Potato Chip bag. Available as a mail order premium exclusively through Fairmont Potato Chips. **$15-25**

ASTRO BOY

In October of 1964, the popularity of Astro Boy reached its greatest impact on America. The action-packed Japanese-produced cartoon surpassed all three national network stations in the ratings. Astro Boy received higher ratings than Maverick, Cheyenne and Superman. In New York, more people tuned in to Astro Boy every night than the network news. Astro Boy was originally a comic book character, created by Osamu Tezuka, one of Japan's leading cartoonists, in the early 1950's. The first cartoon was produced in 1960 by Mushi Productions -Video Promotions.

13

14

13) ASTRO BOY CHILDREN'S RECORD (Golden 1964) 45 rpm record comes in illustrated paper sleeve and contains theme song. **$10-15**

14) ASTRO BOY RECORD ALBUM (Golden 1964) 33-1/3 rpm record contains theme song and stories. **$25-35**

15

15) ASTRO BOY TARGET SET (Atom 1960) Japanese made/distributed. 14"x8"x3"deep box contains six die-cut cardboard stand-ups of Astro Boy and characters and metal air pump rifle which shoots corks. **$200-300**

16

16) ASTRO BOY TATTOO WRAPPER (Topps 1960's) 1.5"x3.5"tattoo wrapper features portrait of Astro Boy on wrapper. Tattoo is on reverse side. **$15-25**

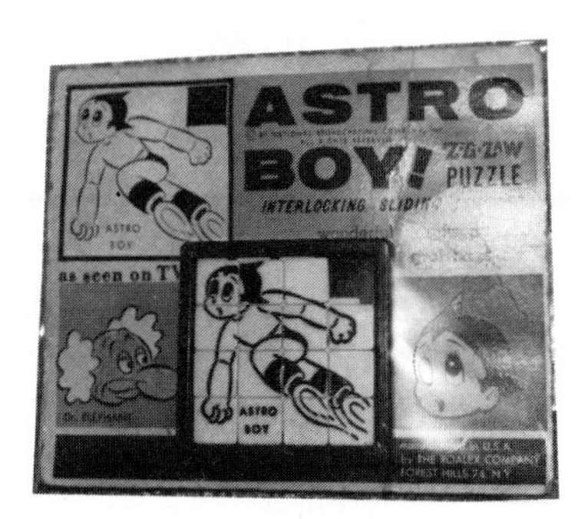

17

17) ASTRO BOY TILE PUZZLE (Roalex 1960's) 5"x6" display card holds black/white plastic puzzle with movable tiles to form image of Astro Boy. **$25-50**

18

18) ASTRO BOY TIN LITHO FRICTION DRIVE SPACE CAR (Modern Toys 1950's) Slick 12" colorfully illustrated tin litho space-like car with vinyl head of Astro Boy. **$300-400**

19

19) ASTRO BOY TIN FRICTION ATOM PATROL CAR (1960's) 13" long tin litho friction drive sedan with Astro Boy and friends illustrated throughout. **$500-600**

20

20) BEN COOPER HALLOWEEN STORE DISPLAY POSTER FOR 1960 (Ben Cooper 1960) Large illustrated 17"x18"poster with eleven cartoon characters including: Huck Hound, Yogi Bear, Fred Flintstone, Deputy Dawg, Rocky Squirrel, Felix the Cat, Dennis the Menace, Quick Draw McDraw, Heckle & Jeckle, Mickey Mouse and Donald Duck. Opposite side has a complete cartoon and character listing of all the Ben Cooper costumes made for the year. **$40-60**

21

22

21) CARTOON CARNIVAL VIEWMASTER REEL SET (Sawyer 1962) 5"x5"color illustrated photo envelope features Supercar, Alvin and the Chipmunks and King Leonardo on the cover and one reel each inside plus storybook. **$25-50**

22) EASY SHOW PROJECTOR MOVIE WHEELS (Kenner 1969) 5"x7"display card contains snap-in movie wheel containing two separate cartoons.

A) **Moby Dick / Mighty Mightor** **$ 8-12**
B) **Superman / Rocky & Bullwinkle** **$10-15**
C) **Casper / Bozo** **$ 8-10**
D) **Fred Flintstone / Yogi Bear** **$ 8-10**
E) **George of the Jungle/Tom & Jerry** **$10-15**
F) **Popeye / Alvin & the Chipmunks** **$ 8-10**
G) **Bugs Bunny / Porky Pig** **$ 8-10**

CLYDE CRASHCUP

Crazy inventor Clyde Crashcup and his assistant Leonardo appeared in ten-minute cartoon episodes on the Alvin show which ran on CBS from October 1961 to September 1962.

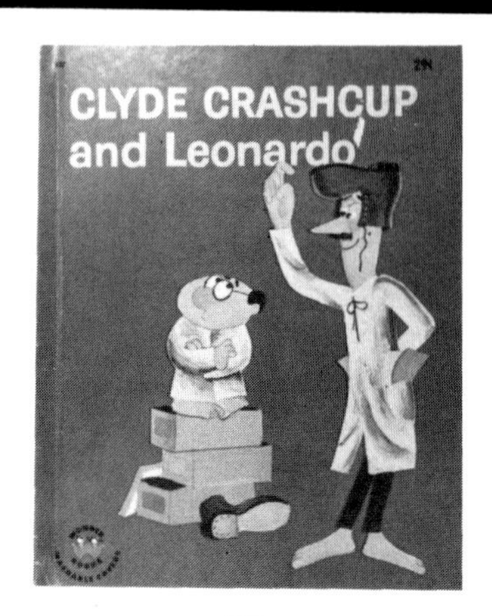

23

23) CLYDE CRASHCUP AND LEONARDO BOOK (Wonder Book 1965) 7"x8"hardback book with 20+ page story with full color story art on each page. **$15-20**

COURAGEOUS CAT

Created by Bob Kane, (of Batman fame), Courageous Cat and sidekick Minute Mouse fight for truth and justice and appeared as a half-hour cartoon show (five minute segments) in 1961. A Trans-Artist Production.

24

24) COURAGEOUS CAT HALLOWEEN COSTUME (Halco 1950's) 10"x12"window display box contains mask and one-piece red/black fabric body suit with yellow print design depicting Courageous Cat shooting ray gun. **$50-75**

25

25) COURAGEOUS CAT TV SOUNDTRACK RECORD ALBUM (Simon Says 1962) Contains story "Around the World in a Daze." **$15-20**

FANTASTIC VOYAGE

This half-hour cartoon program was a spin-off of the popular box-office film of the same name. Released by Filmation Studios for ABC television from 1968 to 1969.

28

28) FANTASTIC VOYAGE BOARD GAME (MB 1968) **$20-25**

DROOPY THE DOG

Created by Tex Avery in 1943 and released through MGM in 1958 as a theater short, Droopy's cartoons later appeared on MGM's Tom and Jerry Show, a half-hour television show which ran from 1965 to 1972.

26

26) DROOPY DOG "POPPING HEAD" FIGURE (General Mills 1960) Plastic 2" tall detailed figure of the cartoon character Droopy the Dog has a spring-loaded head which shoots several feet into the air with a press of a lever. This snap-together assembly toy came free inside General Mills cereals. **$20-25**

27) DROOPY DOG HALLOWEEN COSTUME (Collegeville 1952) One-piece, flannel suit with illustration of Droopy Dog on front. Comes with pullover flannel mask in 10"x12"window box. **$25-50**

THE FUNNY COMPANY

Five-minute episodes revolve around a clubhouse with characters including Buzz and Shrinking Violet. Premiered on ABC from September 1963 to 1965 by Throughout the World Productions.

29 30

29) THE FUNNY COMPANY FRAME TRAY PUZZLE (Whitman 1965) Inlay jigsaw puzzle in a 11"x14"frame tray depicts "Buzz" conducting the gang in a sing-a-long. **$10-15**

30) THE FUNNY COMPANY AND SHY SHRINKIN' VIOLETTE BOOK (Top Top Tales 1965) 10"x8"hardback book with 28-page story accompanied by full color story art on each page. **$8-12**

THE GROOVIE GOOLIES

Count Dracula was one of the many Goolies to occupy Horrible Hall, a Filmation Studios Production, which premiered on CBS in September 1971 to September 1972. Voices included Larry Starch of F-Troop fame. A Filmation Studios Production.

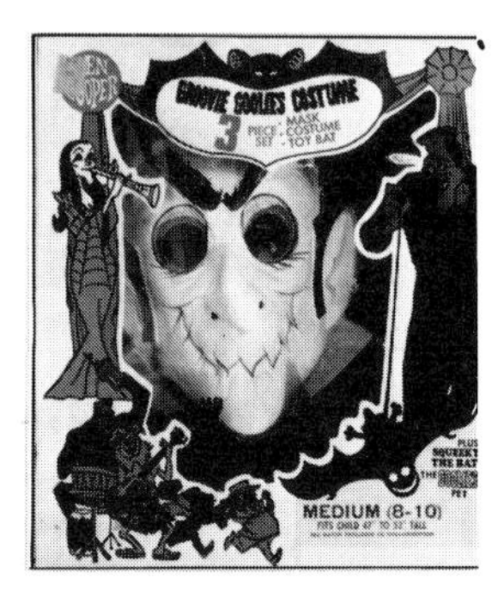

31

31) DRAC "GROOVIE GOOLIES" HALLOWEEN COSTUME (Ben Cooper 1962) Large 11"x12"window display box features color illustration of Groovie Goolies and contains black one-piece rayon fabric outfit with colorful illustration of Drac, plastic mask and 10" long plastic detailed figure of Squeeky the Bat. **$35-60**

KING KONG

Sixty-foot King Kong befriends young Bobby Bond, son of an American scientist stationed on a remote island. Together, they battle the evil Dr. Who, who continually tries to capture Kong for sinister purposes. Premiered on ABC September 1966 to September 1969. A Rankin-Bass Production.

32

32) KING KONG BOARD GAME (MB 1966) **$20-25**

33) KING KONG COLORING BOOK (1966) Has actual felt on Kong's body on the cover. 8"x11",80+ pages. **$15-20**

34

34) KING KONG JUNGLE PLAYSET (Multiple 1967) 10"x12"x5"deep illustrated box contains 8" detailed plastic figure of King Kong with a magnetic hand to hold Bobby Bond figure. Set also includes three painted figures of the Bond family, natives, animals, working animal trap cage, bridge, boat, Skull Island and lots of smaller accessories. **$200-400**

35

35) KING KONG TALKING DOLL (Mattel 1966) Well made 24"tall stuffed plush doll with vinyl Bobby Bond figure. Comes in attractive window display box. **$100-200**

KING LEONARDO AND HIS SHORT SUBJECTS

Thirty-minute cartoon series about a benevolent king lion and his loyal assistant Odie Colognie and their strugglesagainst villains BiggyRat and Itchy Brother. Premiered on NBC September 1960 to 1963. Later syndicated with the title abbreviated to "The King and Odie." Reruns were also used in the half-hour cartoon program Tennessee Tuxedo and his Tales in 1966. A Leonardo Production/Total Television Productions.

36

37

36) KING LEONARDO ART SET (True Talent 1962) 7"x5" box with display card contains eight pre-numbered illustrations, four vials of watercolors, paint brush and instruction sheet. **$35-50**

37) KING LEONARDO COLORING BOOK (Whitman 1961) 8"x11",100-pages. **$20-30**

38) KING LEONARDO DOT TO DOT AND CUT OUT BOOK (Whitman 1962) 8"x11",100+ pages. **$25-35**

39

39) KING LEONARDO HALLOWEEN COSTUME (Collegeville Costumes 1961) 10"x12"box contains mask and one-piece fabric bodysuit, which is red with leopard skin trim around the neck, with an illustration of King Leonardo on front. **$25-50**

40

41

40) KING LEONARDO AND THE ROYAL CONTEST BOOK (Tip Top Tales 1961) 7"x8"20+ page story with full color story art on each page. **$8-12**

41) KING LEONARDO AND HIS SHORT SUBJECTS CHILDREN'S RECORD (Golden Record 1961) 6"x8"stiff paper sleeve contains 78 rpm record containing cartoon theme song. **$12-18**

42

42) KING LEONARDO AND HIS SUBJECTS BOARD GAME (Milton Bradley 1960) **$25-50**

43

43) KING LEONARDO AND HIS SHORT SUBJECTS "BUILD-A-PICTURE" (Jordean 1961) 10"x12"box contains peel-and-stick illustrations of King Leonardo and Odie, mounting boards and other accessories to make two finished wall decorations. **$35-65**

44

44) KING LEONARDO AND HIS SHORT SUBJECTS JIGSAW PUZZLES (Jaymar 1962) 17"x10" box contains interlocking jigsaw puzzle depicting King Leonardo with his subjects. EACH: **$25-40**

LARIAT SAM

Ten-minute cartoon episodes which appeared on the Captain Kangaroo Show 1961 to 1963 on CBS and featured Tippy Toes the Wonder Horse and villains Badlands Meeney and J. Skulking Bushwack. A CBS Production.

45

46

45) LARIAT SAM COLORFORMS SET (1962) Cartoon from Captain Kangaroo. 8"x12"box contains die-cut thin vinyl stick-on character pieces and accessory pieces, plus background scene for placement of characters, and fold-out instruction booklet. **$15-25**

46) LARIAT SAM MAGIC SLATE (Lowe 1962) 8"x12" cardboard display card holds lift-up erasable film sheet and plastic stylus. **$10-20**

47

47) LARIAT SAM PLUSH DOLLS (Rushton 1962) 14" stuffed plush dolls with felt material. Four characters were produced: Lariat Sam, Tippytoes Horse, Badlands Meeney and J. Skulking Bushwack. Our photographs show Lariat Sam and J. Skulking Bushwack. EACH: **$35-65**

LAUREL & HARDY

Produced by Hanna-Barbera and Larry Harmon Pictures, these five-minute episodes used the basic Laurel and Hardy slapstick comedy formula and premiered in 1966. (Harmon was the creative force behind Bozo the Clown.)

48

49

48) LAUREL & HARDY "CHILLER DILLER THRILLER" CHILDREN'S RECORD (Peter Pan 1962) 7"x7" illustrated paper sleeve contains 45 rpm record with songs/stories. **$10-15**

49) LAUREL & HARDY COLORING BOOK (Whitman 1968) 8"x11",80-pages. **$5-10**

50

50) LAUREL & HARDY MAGIC SLATE (1963) 8"x12" cardboard display card holds lift-up erasable film sheet and plastic stylus. **$15-20**

51) LAUREL & HARDY STUFF & LACE DOLLS (TransOgram 1962) Large 12" doll kits comes in 17"x9" illustrated window display box. EACH: **$20-25**

52

53

52) LAUREL PLASTIC FIGURE (Parks 1962) 5" thin plastic hollow figure of Stan Laurel dressed in red/white striped night gown, black shoes and hat with attached surgeon's light. **$25-50**

53) LAUREL VINYL BANK (1972) Large 15" detailed figure bank. **$15-20**

LITTLE LULU

Adapted from Marjorie Buell's comic strip character, animator Max Fleischer brought Lulu to screen life in the 1940's. Twenty-six cartoons were produced between 1944 and 1948 and released through Paramount Pictures.

55

56

54) LITTLE LULU BOARD GAME (MB 1945) **$100-150**

55) LITTLE LULU CHILDREN'S BOOK "LULU'S MAGIC TRICKS" (Little Golden 1954) **$8-15**

56) LITTLE LULU CHILDREN'S RECORD (1950's) Colorful sleeve of Lulu and her kite. **$10-20**

57

58

57) LITTLE LULU COLORING BOOK (Whitman 1946) **$20-30**

58) LITTLE LULU KLEENEX TISSUE BOX (Kimberly-Clark 1956) Colorful 9" stiff paper stand-up of Little Lulu dressed in a red band leader's uniform and holding a box of Kleenex tissue. Punch-out holes in chest to place an accordion-style folded Kleenex to give Lulu a ruffled white blouse. **$10-20**

59

60

61

59) LITTLE LULU COLORING BOOK (Whitman 1968) 8"x11",80+ pages. **$8-10**

60) LITTLE LULU VINYL FIGURE BANK (late 1960's) 8" figure of Lulu next to a fire hydrant. **$20-25**

61) TUBBY PLAYING CARDS (Western Publ. 1960s) Little Lulu's friend Tubby is on the back of this deck of miniature playing cards. Comes in illustrated box. **$5-10**

THE MIGHTY HERCULES

Based on the mythological Greek god Hercules, who from high upon Mount Olympus watches over the Learien Valley of Greece, protecting all from long-time nemesis Daedalus. Released through Adventure Cartoon Productions in 1963.

62

62) THE MIGHTY HERCULES COLORING BOOKS (Lowe 1964) 8"x11"

- **A)** 10 cent cover, 30+ pages. **$15-20**
- **B)** 29 cent cover, 80+ pages. **$25-35**

63

63) MIGHTY HERCULES GAME (Hasbro 1963) Hasbro as a rule made few board games, and those that were produced were small, simple games that sold for under $1.00. Hasbro spared no expense however, when it produced the Mighty Hercules game, which is equal to the quality of Transogram and Ideal deluxe-made games. Box is 10"x20"and contains four cardboard stand-up figures of Hercules, ten illustrated villain cards, ten illustrated Hercules cards and playing board. Object of the game is to be the first player to avoid the clutches of Hydra and Wilamene and safely reach the top of Mount Olympus. **$100-150**

64

65

64) MIGHTY HERCULES MAGIC SLATE (Lowe 1963) 8"x11"illustrated cardboard display card holds lift-up erasable film sheet and comes with wood stylus. **$20-35**

65) MIGHTY HERCULES TATTOO WRAPPER (Fleer Corp. 1961) 1.5"x3.5"tattoo wrapper features portrait of Might Hercules on wrapper. Tattoo is featured on reverse side. **$25-35**

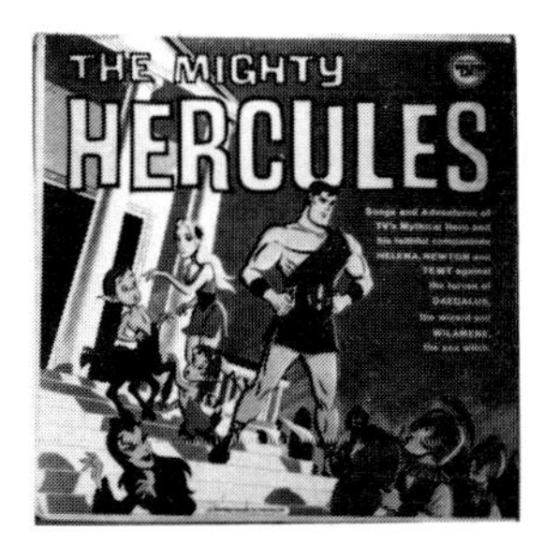

66

66) THE MIGHTY HERCULES RECORD ALBUM (Golden Records 1964) Theme song and stories. **$25-50**

67 68

67) THE MIGHTY HERCULES STICKER BOOK (Lowe 1963) 9"x12" book consists of paper die-cut stickers of characters which can be applied over pre-printed picture pages. **$15-25**

68) NEWT AND TEWT COLORING BOOK (Lowe 1963) 8"x11",40+ pages. **$15-20**

69

69) MILLIE THE LOVABLE MONSTER COLORING BOOK (Saalfield 1963) 8"x11",80-pages. **$20-30**

MILTON THE MONSTER

A half-hour cartoon program which also featured cartoon episodes of "Fearless Fly." Premiered on ABC October 1965 to 1967. Created by Hal Seegar who also created Bat Fink.

70

70) MILTON THE MONSTER BOARD GAME (MB 1965) 8"x16"box contains playing board, four playing pieces and illustrated spinning dial. Object of the game is to be the first player to get his two playing pieces from start to Horror Hill. **$25-40**

71

71) MILTON THE MONSTER AND FEARLESS FLY TATTOO WRAPPER (Fleer Corp. 1967) 1.5"x3.5"tattoo wrapper features portrait of Milton and the Fearless Fly on wrapper. Tattoo is featured on reverse side. **$35-50**

PINK PANTHER

Originally cartoon shorts for theaters, these popular cartoons were re-run as a half-hour television cartoon show which premiered on NBC in 1969. In 1976, new cartoons were produced for "The Pink Panther Laff and A-Half Hour" show which aired on NBC.

72

72) PINK PANTHER BOARD GAME (MB 1969) 10"x20" box contains playing board, dice, and illustrated playing tiles of the Pink Panther. Object of the game is to safely pass the hidden dangers and be the first one to get both of his Pink Panther playing pieces to the finish line. **$25-40**

73

73) PINK PANTHER BENDABLE FIGURE (Bendy 1970) 12" soft pink vinyl bendable figure. **$35-60**

74

74) PINK PANTHER CAR (Dinky 1969) 7" pink plastic and metal car with 2" figure of the Pink Panther behind wheel. The car's main feature is that it will travel across flat surface by use of a gyroscopic road wheel and rack rod. **$50-100**

75

75) PINK PANTHER "CARTOONARAMA" STUDIO ART SET (Cartoonarama 1970) 14"x10"box contains six acetate cartoon cells, each featuring a different character from the half hour Pink Panther cartoon show, six picture frames, eight vials of paint and brush. **$35-60**

76 77

76) PINK PANTHER COLORING BOOK (Whitman 1975) 8"x11",40-pages. **$5-10**

77) PINK PANTHER FIGURAL WALL HANGING TOOTHBRUSH HOLDER (Avon 1970's)4"x6"illustrated box contains 5" tall pink plastic figure of the Pink Panther marching with a drum. Included are two child size toothbrushes which fit into each hand to create the effect he is holding drumsticks. **$12-18**

78

79

78) PINK PANTHER FIGURE (Dakin 1970) 12" tall pink/white hard plastic figure with movable arms and head. **$15-20**

79) PINK PANTHER & INSPECTOR CERAMIC THIMBLE HEADS (1972) Colorful small detailed ceramic heads. **$20-25**

80) PINK PANTHER HALLOWEEN COSTUME (Kusan 1969) 10"x12"box contains mask and one-piece fabric bodysuit with illustration of Pink Panther on front. **$25-35**

THE NEW ADVENTURES OF PINOCCHIO

Puppet marionettes/animation was the medium when the new Pinocchio debuted in 1961. Produced by Video Crafts, Inc. (Video Crafts also produced the very off-beat Wizard of Oz cartoon show in 1960.)

81

81) THE NEW ADVENTURES OF PINOCCHIO GAME (Lowe 1960) Based on the short-lived puppet animation cartoon, this unique game features remote-control magnetic movement which guides Pinocchio figure across the playing board. Object is to help Pinocchio find the Blue Fairy and turn him into a real boy while avoiding the clutches of the villainous Fox and Cat. Box is 12"x15". **$25-50**

82

83

82) THE NEW ADVENTURES OF PINOCCHIO "CRICKET" VINYL FIGURE (1962) 6" painted vinyl figure depicts Pinnochio's friend, Cricket, sitting on a stump dressed in yellow tuxedo coat, red bow tie and yellow straw hat. Figure has movable arms and head. **$15-25**

83) TV's NEW ADVENTURES OF PINOCCHIO BATTERY-OPERATED FIGURE ON XYLOPHONE (Rosko Co. 1960) 9" tall tin-litho figure with molded vinyl head and hands and felt hair and clothes is seated on a log in front of a xylophone. When operating, figure is designed to move his arms up and down as body moves back and forth playing the xylophone's keys. Figure and xylophone come on 6"x8"tin-litho base. **$75-100**

ROGER RAMJET

A proton energypill gives American Eagle Squadron leader Roger Ramjet the power of twenty atom bombs for twenty seconds. He uses the power to battle the evil Noodle Romanov of N.A.S.T.Y. A Ken Snyder Production premiering on ABC in 1965.

84

85

84) ROGER RAMJET COLORING BOOK (Whitman 1966) 8"x11",120 pages. **$25-30**

85) ROGER RAMJET TRAY PUZZLE (Whitman 1965) 11"x14"frame tray puzzle. **$25-30**

86

86) ROGER RAMJET FUN FIGURES (AMF 1965) Rubber Bendable figures ranging in size from 3" to 5" of **Roger Ramjet, Yank & Doodle, Dan & Dee, Lance Crossfire, Lotta Love, Noodles Romanoff, Solenoid Robot** and **General Brassbottom**. EACH: **$25-45**

SPACE ANGEL

Premiered in 1960 and was syndicated through 1965. A TV Comic Strip, Inc. production.

87

87) **SPACE ANGEL GAME** (Transogram 1965) 10"x17"box. Object is to deliver fleet safely home. **$30-60**

TENNESSEE TUXEDO

The adventures of a street-wise penguin and his walrus friend Chumley crusade to change the living conditions of their Megopolis Zoo and were often aided by their scientific friend Phineas J. Whoopee (voice by Larry Starch of F-Troop fame). The voice of Tennessee was by Get Smart star Don Adams. A Leonardo Production/Total Television Production. Premiering on CBS as a half-hour cartoon show in 1963. Other cartoon segments included the King and Odie and Tooter.

88

89

88) **TENNESSEE TUXEDO BOOK** (Saalfield 1963) "Wreck of old 99" 7"x9"softbound, 30-pages. **$15-20**

89) **TENNESSEE TUXEDO MAGIC SLATE** (Saalfield 1963) 8"x11"cardboard back with plastic draw sheet. **$40**

90

90) **TENNESSEE TUXEDO JIGSAW PUZZLE** (Fairchild 1970) Sixty big pieces in colorful 10"x14"box. **$15-25**

91

91) **TENNESSEE TUXEDO PENCIL BY NUMBER COLORING SET** (TransOgram 1963) 10"x17"window display box contains 12 pre-numbered sketches, six color pencils and sharpener. **$50-100**

92

93

92) **TENNESSEE TUXEDO SCHOOL BAG** (Ardee 1969) 12"x10"plastic vinyl briefcase-style school bag with 5x7"color illustration on front of bag. **$75-100**

93) TENNESSEE TUXEDO "SOAKY" BUBBLE BATH CONTAINER (Colgate-Palmolive 1963) 10" plastic soap container with hard plastic removable head. **$15-30**

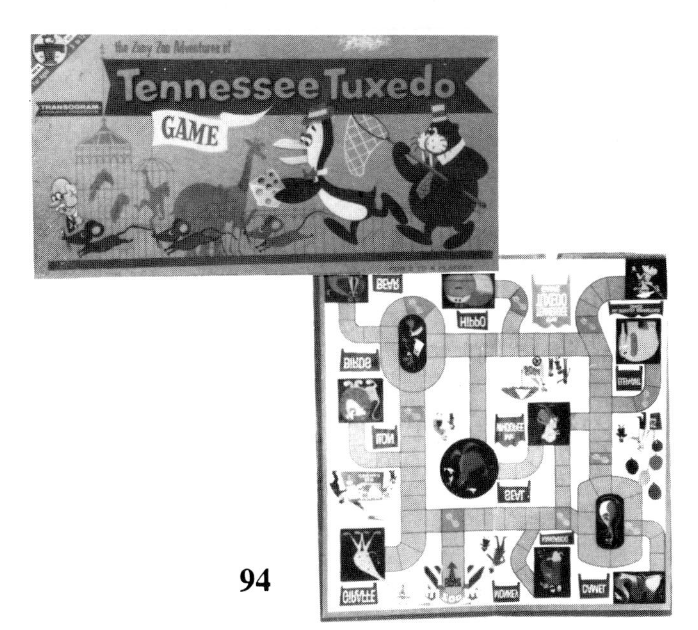

94

94) THE ZANY ZOO ADVENTURES OF TENNESSEE TUXEDO GAME (Transogram 1963) 10"x20"box contains four stand-up figures each of Tennessee, Chumly and the Professor, five cards, 31 animal tiles, spinner and board. Object is to capture mice and zoo animals. **$150-200**

TOM & JERRY

Created in 1940, the feuding Tom Cat and Jerry Mouse won MGM seven Academy Awards during the 1940's. Joseph Hanna and William Barbera produced and directed the cartoons from 1955 to 1958 before leaving MGM to start their own production company. New cartoons were produced for the "The Tom and JerryShow" (CBS, 1965-1972) and "The Tom and Jerry/Grape Ape Show" (ABC, 1975).

95

95) TOM & JERRY BENDABLE FIGURES (Amscan, Inc. 1968) Pair of well detailed figures of Tom and Jerry. Tom is 7" tall, Jerry is 2.5". Figures come in colorfully illustrated 4"x2"x2"display box. **$25-50**

96

96) TOM & JERRY BOARD GAME (Parker Brothers 1948) **$50-75**

97

97) TOM & JERRY BOARD GAME (S&R 1962) 10"x20' box contains playing board, four Jerry mice pieces, four Tom the Cat discs, one spinner and 18 goody discs which consist of cheese, cookie, cake and other food illustrations. Object of the game is to be the first player to safely pass Tom and gather three goodies of any kind. **$25-50**

98

98) **TOM & JERRY BOARD GAME** (TransOgram 1965) **$15-25**

99) **TOM & JERRY BOARD GAME** (MB 1968) **$15-25**

100 101

100) **TOM & JERRY COLORING BOOK** (Watkins-Strathmore 1963) 8"x11",80 pages. **$10-20**

101) **TOM & JERRY COLORING BOOK** (Whitman 1963) 100 pages, 8"x11". **$10-20**

102

102) **TOM & JERRY FIGURE** (Marx 1974) 6" hard plastic figures of Tom and Jerry comes in 6"x9"window display box. **$25-50**

103 104

103) **TOM & JERRY JIGSAW PUZZLE** (Whitman 1963) 8"x10"box contains 60+ piece jigsaw puzzle depicting the Tom and Jerry gang in a tree house. **$12-20**

104) **TOM & JERRY TRAY PUZZLE** (Whitman 1959) Inlay jigsaw puzzle in a 11"x14"frame tray. **$10-15**

105

105) **TOM FIGURAL CIGARETTE HOLDER** (1950's) 7" tall painted glazed china figure of Tom sitting beside a top hat which serves as the holder. **$50-75**

106) **TOM HALLOWEEN COSTUME** (Halco 1952) 10"x12" box contains mask and one-piece fabric bodysuit with illustration of Tom on front. **$25-40**

UNDERDOG

From humble canine "Shoeshine Boy" emerged Underdog, fighter for truth and justice, continually thwarting the evil plots of Simon Bar Sinister and saving his love, Sweet Polly Purebred. Voice of Underdog by Wally Cox. Produced by Leonard Productions/TTP and premiered as a half-hour cartoon show on NBC in October 1964 to 1966. Other cartoon segments included The Hunter (formerly seen on King Leonardo and His Short Subjects) and the Go-Go Gophers.

107

109

107) UNDERDOG COLORING BOOK (Whitman 1965) **$10-20**

108) UNDERDOG DRINKING GLASS (Pepsi 1973) **$10-20**

109) UNDERDOG FIGURAL BANK (1972) 13" tall heavy plastic figural bank in color accents of red, blue, yellow and black. **$25-40**

110

110) UNDERDOG GAME (Milton Bradley 1964) 8"x16"box contains playing board, 24 cards and four playing pieces. Object of the game is to be the first player to save Sweet Polly Purebread from the evil Simon Bar Sinister. **$20-30**

111

111) UNDERDOG HARMONICA (Lednardo 1975) 8" long yellow/plastic child's harmonica with well detailed embossed figures of Underdog and Simon Bar Sinister on each end. Harmonica reads "Have no fear...Underdog is here!" with the word "Underdog" in actual logo style lettering. **$10-20**

112

113

112) UNDERDOG JIGSAW PUZZLE (Whitman 1969) 10"x12"box contains 100-piece puzzle which when assembled creates a 14"x18" scene of Underdog and Sweet Polly Purebred. **$15-25**

113) UNDERDOG FRAME TRAY PUZZLE (Whitman 1965) 11"x14"frame tray puzzle features Underdog saving Sweet Polly Purebread from the clutches of the evil Simon Bar Sinister. **$12-20**

114

115

114) UNDERDOG TATTOO WRAPPER (Fleer Corp. 1966) 1.5"x3.5"tattoo wrapper features portrait of Underdog on wrapper. Tattoo on reverse side features one of many different Jay Ward characters. **$25-35**

115) UNDERDOG SIMON BAR SINISTER DRINKING GLASS (1972) **$15-25**

WINKY DINK

A unique and innovative half-hour cartoon series which enabled children actually to participate in the cartoon. By attaching a transparent sheet to the television screen, kids could assist Winky in escaping from danger by drawing on the sheet (a bridge to cross a cliff or ladder to escape a pit, etc.) with ordinary crayons. The official Winky Dink Magic TV Kit, available through CBS, contains magic screen, crayons, activity book and accessories and at 50 cents, sold millions. Premiering on October 1953, "Winky Dink and You" was hosted by Jack Barry and ran for four and a half years. In 1969, Winky reappeared briefly in a five-minute syndicated cartoon. An Ariel Production.

116

116) WINKY DINK KIT (Standard Toykraft 1954) 12"x15"x1"box contains plastic magic screen and materials for the child to actually participate in Winky's adventure by placing the magic screen over the television screen and drawing pictures when needed (e.g. draw a ladder to help Winky out of a pit, etc.). Materials included illustrated boxed crayon set, illustrated erasing mitt, large 12"x12"TV game book, small jigsaw puzzle and 40 thin vinyl stick-on "doodles." Box lid has display window shaped like a television screen. **$50-100**

117

117) WINKY DINK CLAY DOODLE (1950's) Long 20"x6"x2"deep box contains various colors of children's molding clay. **$25-50**

118

118) WINKY DINK JIGSAW PUZZLE (Jaymar 1950's) 6"x8"box contains 60+ piece jigsaw puzzle of Winky Dink capturing pirate gang. **$25-50**

119

120

119) WINKY DINK LITTLE GOLDEN BOOK (Golden Press 1956) 7"x8"hardback book with 20+ pages with full color story art accompanying each page. **$10-15**

120) WINKY DINK PAINT SET (Pressman 1950's) 18"x15" box contains pre-numbered sketches, water paint tiles, crayons, brush and water tray. **$50-75**

121

121) WINKY DINK TRAY PUZZLE (1950's) Inlay jigsaw puzzle in a 11"x14"frame tray. **$25-50**

WIZARD OF OZ

Video Crafts, Inc., a new wave, very off-beat cartoon production company, took classic literature characters and brought them to cartoon form. (Video Crafts, Inc. also produced "The New Adventures of Pinocchio" in 1961-1962). The Wizard of Oz premiered in 1960 to 1962.

122

122) WIZARD OF OZ GAME (Lowell 1962) Object of the game of course is to follow the yellow brick road to see the Wizard. Box is 12"x15". **$35-50**

123) WIZARD OF OZ HAND PUPPETS (1962) 11" handpuppets of Scarecrow, Lion, Tinman, Dorothy and Toto with molded vinyl heads and fabric bodies. EACH: **$15-20**

124

124) WIZARD OF OZ OIL PAINTING BY NUMBERS SETS (Hasbro 1967) Window display box contains ten vials of paint, brush and twelve pre-numbered sketches.

A) 10"x12"box **$15-25**
B) 14"x16"box **$25-35**

125

125) WIZARD OF OZ STUFFED DOLLS (Artistic Toy Co. 1962) 14"stuffed dolls with cloth outfits and soft molded vinyl heads. Our photo shows the Wizard, cowardly lion and Dorothy. **Each: $35-75**

127

126

126) WIZARD OF OZ TALKING HANDPUPPET (Mattel 1967) 10" talking hand puppet featuring a vinyl head of Dorothy, the Tin Man, Scarecrow and the Wizard. The hand cover is fabric and features the Cowardly Lion. There is a hidden voice box in the hand cover with a pull-string which, when activated, produces eleven different phrases. **$35-50**

127) TALES OF THE WIZARD OF OZ COLORING BOOK (Whitman 962) 8"x11",80+ pages. **$15-25**

BIBLIOGRAPHY

Aldens Mail Order Catalog, 1946-1970

Brooks, Tim & Marsh, Earle. The Complete Directory to Prime Time Network TV Shows, 1946-Present. (Revised Edition) New York: Ballantine Books, 1988.

Bruegman, Bill. Aurora History and Price Guide. Akron, OH: Cap'n Penny Productions, 1992.

Bruegman, Bill. *Toy Scouts, Inc. Mail Order Catalog*, 1983-1993.

Bruegman, Bill. Toys of the Sixties, a Pictorial Price Guide. (Third Printing) Akron, OH: Cap'n Penny Productions, 1992.

Grossman, Gary H. Saturday Morning TV. New York: Dell Publishing Co., Inc., 1981.

Hasbro Product Catalog, 1960-1969.

Ideal Toy Company Product Catalog, 1960-1965.

JC Penny Company Mail Order Catalog, 1946-1970.

Lenburg, Jeff. The Encyclopedia of Animated Cartoon Series. Westpoint, CT: Arlington House Publishers, 1981.

Mattel, Inc. Product Catalog, 1960-1969.

MODEL AND TOY COLLECTOR MAGAZINE, issues 1-23, 1986-1993.

Montgomery Ward Mail Order Catalog, 1946-1970.

PLAYTHINGS, The National Magazine of the Toy Trade, 1948-1971.

Remco Industries Product Catalog, 1960-1969.

Sears Roebuck & Co. Mail Order Catalog, 1946-1970.

Spiegel Mail Order Catalog, 1946-1970.

Standard ToyKraft Company Product Catalog, 1960-1967.

Transogram Toy Company Product Catalog, 1960-1965.

TV JUNIOR, The Children's Television Magazine, 1954-1963.

INDEX

ABOUT THE AUTHOR

In just ten years, Bill Bruegman has become an internationally recognized leader in the profitable hobby of collecting memorabilia. Starting as a dealer in vintage records, Bill soon noticed many of his customers wanted music and collectibles from the TV cartoons of their childhoods.

Spotting a trend toward 1950's and 1960's memorabilia, Bill turned to selling and trading model kits, board games and other toys which led to the formation of his mail-order firm, **Toy Scouts, Inc.**, in 1983. Since that time, Toy Scouts has grown to supply customers around the world.

To better serve his clientele, Bill created in 1986 **Model and Toy Collector Magazine**, a quarterly publication devoted to Fifties and Sixties collectibles, and watched it skyrocket in popularity among hobbyists and investors.

Because of his expertise, Bill has been recognized in leading news sources including Rolling Stone, USA Today, The Wall Street Journal, Entrepreneur, and Pronto, the prestigious Japanese trend monitor. He has also been sought out by motion picture studios such as Warner Brothers and Columbia Pictures as a source for the purchase and rental of period props.

Items from the Toy Scouts inventory and Bill's personal collection were recently on display at the Smithsonian Institute in Washington, DC as part of "It's Your Childhood, Charlie Brown," an exhibit of post-World War II children's life. Other museums, including the Henry Ford Museum and the "Please Touch" Museum for children in New York, have also featured pieces of the Toy Scouts inventory in their displays.

QUISP
POP TARTS
SMOKE
M. LUCK '90
ALL CHARACTERS ARE COPYRIGHT ©1990 ALL RIGHTS RESERVED